Sound: A Very Short Introduction

VERY SHORT INTRODUCTIONS are for anyone wanting a stimulating and accessible way into a new subject. They are written by experts, and have been translated into more than 40 different languages.

The Series began in 1995, and now covers a wide variety of topics in every discipline. The VSI library now contains over 450 volumes—a Very Short Introduction to everything from Psychology and Philosophy of Science to American History and Relativity—and continues to grow in every subject area.

Very Short Introductions available now:

For more information visit our website
www.oup.com/vsi/

Mike Goldsmith

SOUND

A Very Short Introduction

OXFORD

UNIVERSITY PRESS

Great Clarendon Street, Oxford, OX2 6DP,
United Kingdom

Oxford University Press is a department of the University of Oxford.
It furthers the University's objective of excellence in research, scholarship,
and education by publishing worldwide. Oxford is a registered trade mark of
Oxford University Press in the UK and in certain other countries

Published in the United States of America by Oxford University Press
198 Madison Avenue, New York, NY 10016, United States of America

British Library Cataloguing in Publication Data
Data available

Library of Congress Control Number: 2015945427

ISBN 978-0-19-870844-5

Printed and bound by
CPI Group (UK) Ltd, Croydon, CR0 4YY

With thanks to
Richard Barham
Graham Carter
Mark Hodnett
Stephen Robinson
Hilary Whittle

Contents

List of illustrations

Sound

The sound spectrum

	Frequency in hertz
Low-frequency volcanic activity	0.01
Airborne infrasound from sea waves	0.20
Lower limit of elephant hearing	16
Minimum audible tone	20
Lowest piano key	27.5
Lowest recorded sung note	41
Middle C	261.626
Piano key A above middle C	440.000
A above middle C on Bach's organ	480
Highest recorded sung note	1,865
First part of blackbird song	2,500
Highest piano key	4,186.01
highest audible tone, aged 80	8,000
Highest audible tone, aged 10	20,000
Dental cleaning	25,000
Typical ultrasonic welder	30,000
Ultrasonic parking sensor	40,000
Upper limit of dog hearing	60,000
Typical bat scan frequency	80,000
Upper limit of bat hearing	110,000
Upper limit of beluga whale hearing	123,000
Typical fetal scan	15,000,000
Acoustic microscope	8,000,000,000

Chapter 1
Past sounds

Sound: 13.7 billion years ago

Sound has its origin far back in time, not long after the disappointingly silent Big Bang. In fact, sound waves formed as soon as there was a medium for them, which was 300,000 years after the beginning of everything, and thirteen billion years before there was anyone to listen.

The primordial sound was of a very low frequency, but powerful and omnipresent, and it formed when the plasma of the newborn universe arranged itself in pseudo-regular patterns through space. Galaxies eventually formed in the denser regions, including the progenitor of the world we live on and the sun we orbit.

Moving on a few billion years to the first days of Earth (about 4.6 billion years ago), there were sounds aplenty, passing through the solid planet's crust and its subsurface liquid areas, and bouncing and bending through its atmosphere. Eventually, the hot land cooled, rains fell, and oceans formed—oceans full of sound. So, the environment in which the first living things evolved was an acoustically rich one, which profoundly affected the forms, habits, and destinies of those creatures.

Hearing: 500 million years ago

For us, there is a clear distinction between hearing and feeling, but it's a very different matter for undersea creatures: sounds pass as easily through their bodies as vibrations do through ours, and, in fish, are detected by structures called neuromats, which are distributed over their body surface (fish have several other hearing structures too).

Neuromats contain hair cells similar to those in our ears and provide their owner with information about the strengths and directions of local sounds. They evolved perhaps 500 million years ago. To detect airborne sound, eardrums and cochleae are essential, and hence evolved once amphibians began to colonize the land, around 400 million years ago. Communication was probably the main spur to the evolution of hearing, since sounds have overwhelming advantages over visual signals: while the ability to make lights and change colours exists in some marine organisms, putting on a light show is far more challenging and narrower in range than making noise. Noises are made easily—as easily as breathing. In humans, breath-made sounds (precisely controlled by our big brains) gave us the power of speech.

Music: 40,000 years ago

The appreciation of music is a mysterious pleasure and an ancient one too: over 40,000 years ago Neanderthals probably had flutes and, by the time *Homo sapiens* emerged, no well-appointed cave dwelling was complete without a rock gong. It may be that the first humans sang; perhaps even before speech evolved. But why? Enjoying music has no obvious evolutionary advantage. Darwin himself was baffled, but suggested that a taste for music might arise through sounds made in mating rituals, and many echo this view today. Others, however, prefer the suggestion of evolutionary psychologist Steven Pinker that music is an auditory equivalent of cheesecake, which we enjoy not because that preference assisted our ancestors in surviving, but because many of the sensations cheesecake elicits are of evolutionary value in themselves: the sweetness of fruit announces its ripeness and creamy flavours suggest energy-rich fats. Or perhaps music reminds us of birdsong—the presence of which indicates that no large predators are around.

Harmony: 2,500 years ago

Today, sound plays a huge variety of roles in our lives. Many of our inventions are dedicated to its creation, transmission, storage, modification, or reproduction. But the conquest of sound is by no means a recent development: some of the most ancient artefacts we know are musical instruments, and acoustics was one of the first sciences; in 500 BCE or so, Pythagoras discovered that the sound made by a strummed string mingles pleasantly with one made when the string's length is halved. The 'distance' between the two sounds is an octave, by definition and by universal agreement the most harmonious of all pairs of different notes. Sounds almost as harmonious result if the string lengths bear other simple numerical ratios: if one string is one-and-a-half times the other, a *fifth* is produced, for example.

According to legend, Pythagoras made this discovery when he heard tuneful hammer sounds emerging from a forge where a number of blacksmiths were at work. When (being a budding scientist) he weighed the hammers they were using, he found that those pairs which made a pleasant sound had weights which were simple multiples of each other. The fact that this story is told to this day is surprising given that the frequency at which a hammer sounds is *not* fixed by its weight. Whatever actually piqued his interest, the instrument Pythagoras used to study harmony was the monochord—a device with a single string whose length can be set by a moveable bridge.

To Pythagoras, the fact that the pleasantness of sounds was defined through whole-number ratios suggested that numbers were the key to the universe. 'All', so he is supposed to have said, 'is number'. Today's scientists would agree, and, in its impact on scientific method, mathematics, music making, and acoustics, his discovery may be one of the greatest breakthroughs of all.

Although anyone who makes or plays a stringed instrument knows that tension, as well as length, affects the note a string makes (otherwise turning pegs to tune stringed instruments wouldn't work), this was not quantified until the 16th century by Vincenzo Galileo, father of the scientist, who showed that pitch increases with the square root of the tension. We now know that it also depends on the string's thickness and density.

The Greeks were interested in the practicalities of sound, thanks to their keen interest in making their voices heard: plays, orations, declamations, debates, songs, chants, and proclamations abounded. Perhaps their greatest acoustic structure is the theatre of Epidaurus, built in the 4th century BCE. The distance from stage centre to the back row is about 60 metres, yet actors can be heard clearly from any of the 1,400 seats: fifty-five rows in total. It is in these seats that the theatre's acoustic secret lies: the limestone of which they are made, their corrugated surfaces, and the spaces between them all contribute to absorbing sub-500 Hz frequencies and reflecting higher ones, quietening crowd murmur and enhancing performances respectively.

But there are disadvantages to open air speech, even in Epidaurus. With no ceiling to contain the sound, speakers must be very loud indeed, which is not only tiring but also tends to rob the voice of its subtlety. (Despite legends to the contrary, the Greeks lacked megaphones, which were not invented until the 1670s—simultaneously by Athanasius Kircher in Germany and Samuel Morland in England.) Background noise becomes far more intrusive too, though in the Greek theatres performance days would have been quiet, since the majority of the local population would be in the theatre.

Indoor public spaces solved these problems but introduced new ones: echo and reverberation. For an echo to *be* an echo, it must be heard more than about 1/20 of a second after the sound itself. If heard before that, the ear responds as if to a single, louder, sound. Thus 1/20 second is the auditory equivalent to the 1/5 of a second that our eyes need to see a changing thing as two separate images. (Hence, when camera frames move faster than this, we obtain the illusion of flowing movement. This 'persistence of vision' is what makes us see the rapid sequence of still images that make up a cinema film as a smoothly changing image.)

Since airborne sounds travel about 10 metres in 1/20 second, rooms larger than this (in any dimension) are echo chambers waiting to happen. Luckily echo can be reduced by covering hard surfaces with soft, fabric-covered objects, such as audience members.

There is of course more to sound than science or entertainment. Even wordless sounds come freighted with meaning, much of which was attached in prehistoric times. The lonely howl of the wind, a shocking scream of pain, the joyous songs of birds, the happiness of children's laughter: in these and many other cases, evolution has forged unbreakable links between sound and emotion. These emotional attachments have been exploited by us since ancient times: war cries, for example, have long been produced both to chill the blood of the enemy and to unite and rouse the courage of the attackers.

The modern world: acoustics and more

Following the work of the ancient Greeks, little research on the nature of sound was carried out until the 17th century, when Robert Hooke proved by simple demonstrations that frequency and pitch are linked. Although Isaac Newton suggested an equation for the velocity of sound, it was incorrect, and an accurate version was only derived by Pierre Simon Laplace in

> **Box 1**
>
> Frequency of a stretched string $f = \frac{1}{2l}\sqrt{\frac{T}{\mu}}$; length l, tension T, density μ (mu).
>
> Velocity of sound v (or occasionally c from the Latin *celeritas*) $= \sqrt{K/\rho}$; elasticity K, density ρ (rho).

1816. Laplace showed that, to an excellent approximation, the velocity of sound depends only on the density and the elasticity of the medium through which it passes (Box 1).

The invention of the first electroacoustic devices in the mid-19th century led to revolutions in both the understanding and the control of sound: the microphone, telephone, and loudspeaker appeared in quick succession, spurring rapid developments in research, commerce, and the arts.

The 20th century and electronic engineering began together, with the inventions of the diode (the first rectifier, used initially for detecting radio signals) in 1903 and the triode (the first amplifier) in 1906. The development of electronics was greatly accelerated by the World Wars, which also led to the birth of underwater acoustics research through interest in submarine warfare and ship detection.

While the existence of sounds with frequencies too high to hear had occasionally been discussed in the 19th century, they were only investigated in the context of the upper frequency limit of human hearing. Even though such sounds could be made easily by sparks, whistles, air jets, or piezoelectric crystals, they remained of little interest until World War I, when it was realized that they might be pressed into service as part of what we would now call sonar systems. Soon after the war, further investigations revealed that such sounds had a range of unique properties, not all of which could be explained: they could kill living things, cause chemical changes, generate light and heat, and make wood explode in showers of sparks. It may be that they were called supersounds in part because of their strange powers. Although far less famous than X-rays and radium emanations, supersounds soon gained a similar reputation as secret forces wrested from nature by the tools of science, but still retaining a glamour of almost supernatural mystery and power.

It was the middle of the 20th century before the power of ultrasound was properly understood and exploited. By then, the use of amplifiable electroacoustic technology had truly transformed the world. Public speakers, formerly limited to audiences of a couple of thousand at the very most by the sounds of their voices and the sizes of their venues, could now talk to people in their millions, thousands of miles away—either instantaneously or a day or a century later. Life could be captured, recorded, and analysed as never before.

There were whole new fields too, including sound art; an ill-defined discipline with origins in the futurist movement of the 1900s to 1930s, and in the development of electronic music and good recording technology. Luigi Russolo's *Gran Concerto Futuristico* (1917) is an important early example, and a recent one is *Lowlands* by Susan Philipsz, a set of variations on a lament played over one another, which won the Turner Prize in 2010. A related area is ambient music (often called muzak by its detractors), which is designed to provide an appropriate background to public spaces. Brian Eno's *Ambient 1: Music for Airports* (1978) is a good example, and saccharine carols looped endlessly in supermarkets (accompanied by gloomy employees forced to dress as elves) is a bad one.

Ambient music is an example of an artificial soundscape, a concept popularized by Murray Schafer, who helped set up the World Soundscape Project in Vancouver in the late 1960s. The project led in turn to the formation of the World Forum for Acoustic Ecology in 1993. Schafer has been highly influential, in part through his galvanizing claim that acoustic environments can not only reveal the social conditions of those who inhabit them, but even predict how that society will evolve.

Applying Schafer's approach, economist and polymath Jacques Attali argues that changes in musical convention prefigure wider changes in society. Some have gone much farther than this; historian Alain Corbin argues that village bell ringing in 18th- and 19th-century French villages moulded social and economic relations there, and artist and writer Brandon Labelle says that 'my feeling is that an entire history and culture can be found within a single sound'.

More generally, the concept of the soundscape has enjoyed great popularity in a range of disciplines, though Schafer's definition of the term has been extended to take account of the relative and dynamic nature of an acoustic environment: technology historian Emily Thompson points out that it is 'simultaneously a physical environment and a way of perceiving that environment'.

In film, the construction of artificial soundscapes is achieved partly through sound effects. These have also been a mainstay of radio drama since its inception and, in the form of artificial thunder for example, have been heard in theatres since ancient Greece. In films, the craft of designing, producing, and synchronizing sound effects with events on screen is known as Foley.

No longer dependent on our ears to detect sounds or on voices and mechanical devices to make them, we can study and use sounds too low pitched, high pitched, or quiet to hear, and we can generate and direct acoustic beams with enormous power and precision, leading to applications in medicine, defence, mapping, and many other fields. Since World War II, serious attempts have been made to develop sound-based weapons, mainly by producing and directing extremely high-intensity beams. Perhaps the best-known example in current use is the Long Range Acoustic Device (LRAD), which projects either commands or unpleasant sounds. It has been used against humans and wildlife in several countries.

Alongside the deliberate and controlled development of sound, noise pollution has also spread through much of the world. The highly sensitive hearing systems that served our ancestors so well, and still allow our pleasure in music and our facility in speech, are now the conduits of annoyance, stress, and damage.

So while we have gained mastery of the production of sound, we are far from being able to control it. In order to have any chance of doing so, we need to understand its nature.

Chapter 2
The nature of sound

Two faces of sound

If a tree falls in a forest without anyone to hear it, does it make a sound? The dual meaning of 'sound', as physical phenomenon and sensation, provides a clear answer: yes and no. The relationships between the physical and sensual aspects of sound are complex, in that many of the impressions sound makes on us are related to its physical parameters but not reducible to them. So: high-frequency sounds sound higher pitched—usually. And more powerful sounds sound louder—on the whole. Furthermore, many sounds, from sirens to skirls and from lullabies to lions' roars, make emotional impacts on us which have only the vaguest relationships to their physical parameters.

The physical aspects of sound are far better understood than the emotional ones, so it is with physics that we should begin.

Pressure waves

Sounds are usually made by something moving in a cyclic manner: the diaphragm of a loudspeaker pulsing in and out, the gap between the vocal folds narrowing and widening, or a guitar string vibrating back and forth. It is the transmission of these motions to the surrounding medium (solid, liquid, or gas), and their progression through that medium, that constitute sound. In some cases the motion begins in the medium itself, such as the air in the neck of a bottle when one blows across it. Non-moving sources include sudden releases of heat energy, such as by explosions or sparks, and rapidly oscillating heat sources.

When the motion is that of a loudspeaker's diaphragm, the cause is a varying electrical signal with the same pattern as the sound wave that the diaphragm will produce. Each time the diaphragm moves out, it squeezes the air molecules immediately in front of it closer together, forming a region of high pressure. These molecules press on their neighbours, moving them closer together in their turn, and so a pulse of close-together molecules (a *compression*) moves through the medium, followed by a low-pressure area (*rarefaction*), which is produced as the diaphragm moves inwards.

The diaphragm then moves out again, making a second pulse. How often the diaphragm moves in and out during 1 second gives the frequency of the sound wave (in hertz, abbreviated Hz). The simplest sound wave is a pure tone as made, for example, by a tuning fork; a snapshot of the variation in air pressure with distance from such a fork would be a sine wave, as shown in Figure 1.

The distance between adjacent peaks (or troughs) of a sound wave gives the wavelength (λ). The sound will travel through the air at a velocity v, which will be around 340 metres per second at room temperature. The frequency (f) is given by the equation $f = v/\lambda$. A plot of the variation of pressure over time at a single point in

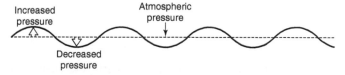

1. **Sound wave pressure plot.**

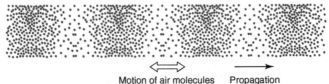

Motion of air molecules   Propagation
associated with sound   of sound

2. Molecular view of sound wave.

space is also a sine wave, so we could label the above x-axis of Figure 1 as 'time' if we wished.

Images like Figure 1 are so commonplace that it is easy to imagine they provide some kind of a picture of a sound wave, and many books use them in this way. In fact, however, there is no up and down (*transverse*) motion in a sound wave as there is in, say, an ocean wave—the only motion is of molecules shuttling alternately away from and towards the source, like the balls of a Newton's cradle. Such waves are referred to as *longitudinal*, and if we could see air molecules they would look something like Figure 2.

If a continuous sound originates from a point then it spreads in all directions, as an expanding sphere. If the detection area is small (like a microphone diaphragm or eardrum) and is several metres from the source, the curvature of the sound sphere is negligible, in which case the sound arrives in the form of plane waves. Even if the source of the sound has a particular direction to it (like most loudspeakers), the sound will still spread spherically, so long as the diaphragm is larger than the wavelength of the sound. For shorter wavelengths, the sound retains its original direction to some extent, and at sufficiently high frequencies will form a beam (see Chapter 6).

Carrying sound

The velocity of sound depends only on the elasticity and density of the medium (Chapter 1). In air, sound velocity increases with

Table 1. Sound velocities in different media and conditions

Material	Velocity of sound (metres per second)	Velocity of sound compared to that in air at 20°C, 80% relative humidity
Hydrogen at 0°C	1,286	3.73
Air at 0°C, 0% relative humidity	331.45	0.962
Air at 0°C, 80% relative humidity	331.70	0.963
Air at 20°C, 80% relative humidity	344.37	1
Water at 25°C	1,493	4.33
Seawater at 25°C	1,533	4.45
Diamond	12,000	34.85
Granite	5,950	17.28
Iron	5,130	14.90
Vulcanized rubber	55	0.16

increasing humidity and decreases with increasing temperature, but only because of the changes in density that these factors cause. Some examples are given in Table 1.

Because the velocity of sound in air increases with temperature, on days when the air many metres up is hotter than that near the ground, sound waves travel faster the higher they are. The effect of this velocity increase is to bend (*refract*) them downwards from the warmer air, returning to Earth some distance away, as shown in Figure 3. Sometimes sounds can be heard more clearly at a great distance than at a short one due to this effect. Refraction also explains why it is hard to hear against the wind: the wind near the ground slows the sound waves slightly (compared to the ground) but the air a few metres up is faster, so the sound waves are slowed a little more there. The sound refracts from the low-velocity region to the higher-velocity one, and hence curves up away from the ground and from your ears (see Figure 4).

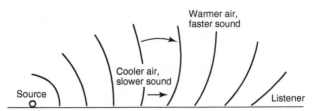

3. Sound propagation when air is cooler near the ground than high above it.

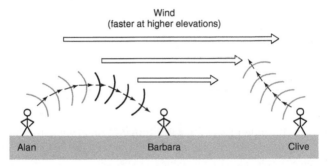

4. Barbara can hear Alan, but not Clive.

No matter what one does with a diaphragm, one cannot push sound through the air that surrounds it any faster. More rapid waggling generates pressure pulses that are closer together, which then arrive at a point—your eardrum, say—more frequently. That is to say, the sound frequency would rise. If one tries pushing the air harder by moving the diaphragm further in and out, then the amount of compression (and rarefaction) in the pulses increases, leading to a higher sound pressure (heard as a louder sound). If one forces the diaphragm to move faster than the velocity of sound in the medium, a pulse has no time to move away from the diaphragm before the next pulse forms. Hence, they pile up into a single, extreme high-pressure pulse known as a shock wave (the cause of sonic booms and whip cracks).

Moving a diaphragm more rapidly is not the only way to increase the frequency of a sound: if the loudspeaker (or other source) is rapidly approaching you or you are rapidly approaching it, the pressure pulses arrive at your ear more frequently, because each starts closer to you than the one before, so the frequency of the sound rises. Once the source has passed by, the pulses hit your ear with longer intervals in between, because each has a little farther to travel than the one before. So the frequency falls. This is the well-known Doppler effect, which happens whenever one is passed by a speeding motorbike or the siren of the police car behind it (Box 2).

Like light, sound reflects, and as with reflection from a mirror, an image of the sound source is formed if a surface is smooth and hard, so if you are somewhere between source and surface, you will hear approximately the same sound from each side (but quieter from the reflector side). 'Smooth' is a relative term though, meaning 'with bumps smaller than the wavelength'. Since sound waves are about one million times longer than light (comparing 3 kHz to yellow), even quite rough surfaces like concrete make good acoustic mirrors. Concave acoustic mirrors focus the sounds that they reflect: in World War I just such concave concrete sound mirrors were built along the south coast of England to focus the sound of approaching planes into the ears of listening solders. When sound echoes between two or more curved reflectors, the result can be a whispering gallery, like the one in London's Saint Paul's Cathedral.

Box 2

Doppler effect: for an approaching sound source, the frequency heard by the observer is $f_{obs} = \left(\dfrac{v+u}{v} \right) f$; velocity of sound v, speed of approach u, frequency of source f. For a receding sound source, the '+' becomes a '−'.

Sound will reflect from the interface between *any* two media, whether air and concrete, water and air, or different rock layers in the Earth. How much of the sound is reflected depends on the difference in the acoustic impedances of the two media, and the impedance in turn depends on the density of the medium and the velocity of sound in it. Acoustic impedance (Box 3) is similar to electrical resistance in that it measures the difficulty with which sound can travel through a medium. It is key to many of the effects and applications of sound. For instance, a soft rubbery surface will absorb sound and convert it to heat, since soft rubber has an extremely high acoustic impedance. Stealth coatings on submarines are based on this fact, but unfortunately the softness of rubber is very temperature-dependent, so, when Cold War submarines were redeployed from the North Atlantic to the Gulf from the late 1980s, the higher water temperatures robbed them of their stealth and set off a flurry of research and resurfacing.

Sound can be focussed by passing it through an acoustic lens, often made of acrylic plastic. Lenses work because a wave is refracted when it passes from one medium to another, so long as it strikes the interface between the media at an angle. The angle through which the wave is refracted depends on the ratio of its velocities in the two media (Snell's law, Box 4).

One effect which is usually far more noticeable for sound than for light is its ability to bend round corners and over walls, and to spread out after passing through an opening, a phenomenon known as diffraction or scattering (Figure 5).

Box 3

Characteristic acoustic impedance of a medium: $Z_0 = \rho_0 v_0$; density ρ_0, velocity of sound v_0. The unit is the Rayl, and the $_0$s indicate that these are the values of the medium when it is 'unperturbed'—that is, when no sound is present in it.

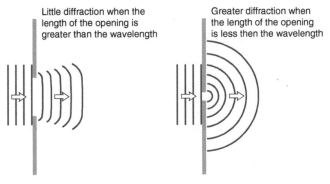

Little diffraction when the length of the opening is greater than the wavelength

Greater diffraction when the length of the opening is less then the wavelength

5. Diffraction.

The longer the wavelength, the greater the bending. So if a high wall is introduced between a sound source—a band, say—and a listener, the bass sounds diffract from its top back down to earth but the high-pitched ones are lost (Figure 6). This muffling effect is a useful clue that helps us gauge the distances of familiar sound sources outdoors.

When light falls on a series of parallel lines, stripes, or ridges a single wavelength apart (or thereabouts), it is diffracted, and since shorter wavelengths are diffracted through larger angles, such *diffraction gratings* split white light into its component colours—the back of a CD makes rainbows from sunbeams in just this way. Since a pure tone is a regular series of 'stripes' of increased pressure, it can also act as a diffraction grating, scattering light with a wavelength around that of the distance between the stripes (that distance being half the wavelength of

17

6. Diffraction of different wavelengths.

the sound). Usually the medium here is a crystalline solid, such as fused quartz. This acousto-optic effect, where sound waves scatter light, is used both underwater and in air as a non-perturbing measurement and imaging tool (see Figure 7).

When sounds from multiple sources meet, mix, and mingle, the result is a three-dimensional pattern of loud and quiet areas called an interference pattern. The quiet areas form at the points where rarefactions from one source meet compressions from another (destructive interference), and the loud ones arise when rarefactions meet rarefactions, or compressions meet compressions (constructive interference) (Figure 8).

Interference is important in stereo sound production and in noise cancellation, and it introduces one more parameter that characterizes a sound wave: its phase, that is, how high or low its pressure is at a particular point in space and time. Phase really only matters when sound waves interact: in the above example,

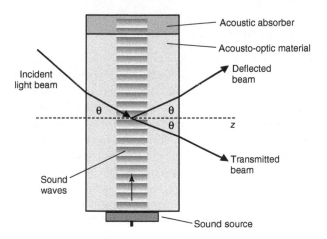

7. **The acousto-optic effect.**

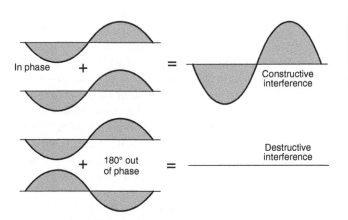

8. **Constructive and destructive interference.**

pairs of sound waves whose compressions coincide with each other (and hence make a loud area) are *in phase*, while those in which they do not coincide are *out of phase*. When waves are maximally out of phase, they are said to be *in antiphase*. Our hearing systems cannot detect phase.

The power of sound

There are several ways in which the amount of sound can be defined and measured, and each way is appropriate for different applications. If hearing or music is the context, *sound pressure* is the obvious choice since it is the parameter which relates most directly (though none too simply: read on!) to the impression of *loudness*. But in discussing the efficiency of a sound source one may wish to know how much energy is flowing from it in a second—the *sound power*. To describe the effects of a particular sound field on an object, the parameter of interest is the *sound intensity*, which is the amount of sound energy striking 1 square metre of that object each second. *Volume* is an ill-defined measure used to label audio equipment, but intended to mimic loudness.

Audible-frequency sound waves lose very little energy through absorption by the air through which they pass (around 0.25 dB/6 per cent per 100 metres, though varying greatly with weather conditions). The main reason sounds die with distance is that they are free to spread out in many directions, so their energies spread progressively more and more thinly to occupy larger and larger volumes. If a sound source is suspended in free air so that its sound can spread in every direction (*spherical spreading*), then the sound pressure is inversely proportional to the distance of the listener from the source: that is, if the distance from source to measurement point doubles, the sound pressure halves.

The intensity of the sound falls more rapidly than this: it is inversely proportional to the *square* of the distance. So, if the distance from source to measurement point doubles, the sound intensity falls to one-quarter ($1/2^2$). If the distance is multiplied by 10, the intensity falls to one-hundredth ($1/10^2$). If the sound source is on the ground, the waves spread hemispherically (Box 5), and sound pressure and intensity fall at half the above rates: in other words the intensity falls to one-half if the distance doubles—roughly,

> **Box 5**
>
> Spherical spreading: $I = P / 4\pi r^2$; intensity I, power P, distance from the source r.

unless the ground is a perfect reflector (a marble floor is pretty near), intensity will fall faster than this, due to loss of energy due to absorption. The sound power depends only on the source, so it is the same at any distance.

Pure tones are never found in nature, but are perhaps most closely approximated by the songs of birds. The waveforms of real sounds look very different: the pressure-plots of different sounds of similar fundamental frequency are shown in Figure 9.

The difficult decibel

Sound was one of the first forms of energy to be understood: as far back as 300 BCE it was known to be some kind of pattern of physical changes that could travel through air and water, but it was long after that that the most obvious characteristic of sound—its loudness—was quantified in any way. What did turn up took over 2,000 years to arrive, and was not very satisfactory when it did.

By far the most widely used way to quantify the amount of sound is the decibel (dB, Box 6): if two signals differ in sound pressure by 1 dB, the ratio of those pressures is about 1.2:1. (Handily, this happens to be about the smallest difference we can hear under ideal conditions.) A 10 dB difference corresponds to a ratio of about 3:1, and a 100 dB difference to a ratio of 100,000:1.

A decibel is one-tenth of a bel, a name made by combining three letters commonly used in transmission theory (β, ε, and l) with a tip of the hat to Alexander Graham Bell. Decibels aren't

Sound

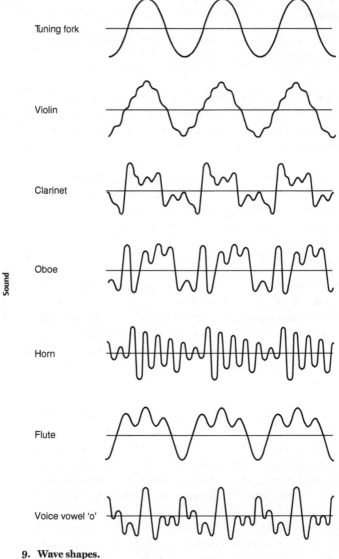

Tuning fork

Violin

Clarinet

Oboe

Horn

Flute

Voice vowel 'o'

9. Wave shapes.

units—they are ratios, so they can describe how much more powerful one thing is than another: you could, if you wished, use them to compare the outputs of a pair of heaters. But that would not tell you anything about how hot either of them actually are.

To describe the sound of a device in decibels, it is vital to know what you are comparing it with. For airborne sound, the comparison is with a sound that is just hearable (corresponding to a pressure of twenty micropascals). When the amount of sound is given in terms of such a reference level, the word 'level' is appended: hence sound pressure level (SPL), for example.

So, a sound of 0 dB is 'one times louder than' (i.e. equally loud as) a sound you can just about hear, 1 dB is about 1:12 times as loud, 2 dB is 1:26 times, and so on. Are all acousticians happy with this solution? No, they are not. Ultrasound engineers don't care how much 'louder than you can just about hear' their ultrasound is, because no one can hear it in the first place. It's *power* they like, and it's watts they measure it in. Meanwhile, underwater acousticians rightly ask: 'The threshold of hearing? What does that mean when your ears are full of water and you've got a rubber headpiece on? Or if you're a whale?' So they base their decibels on a reference pressure of one micropascal, because that's nice and easy to remember. So now we have two kinds of decibel, one for use in water and one for air, which will give different values for the same sound. Not too much of a problem so long as everyone always remembers to say what the reference level of the decibel they are using is. Sadly however, they don't.

There is another problem. Few of us care how much sound an object produces—what we want to know is how loud it will sound. And that depends on how far away the thing is. This may seem obvious, but it means that we can't ever say that the SPL of a car horn *is* 90 dB, only that it has that value at some stated distance. Often, even those handy decibel charts so popular in textbooks get this bit wrong, and claim the SPL of a pneumatic drill is 100 dB, when they really mean '100 dB if measured at a distance of 10 (or however many) metres'. It's not difficult to see where this laziness creeps in—the charts usually also have examples like 'a quiet office', and we understand that the chart maker is referring to a quiet office that you are working in, not a quiet office down the corridor or in another town.

There is a third problem: a source of sound might make sound waves at any one frequency, at a handful or at a wide variety of frequencies. Let's assume for the moment that the source of the sound is a loudspeaker so efficient that it always converts all the electrical energy fed into it into sound. And let's imagine it has a frequency control but no volume knob. If we measured the total sound energy flowing from that loudspeaker each second (that is, the power) while changing its frequency, that power would of course remain constant. Similarly, the SPL at a particular distance from the loudspeaker would stay the same—as a microphone would show (assuming it were equally sensitive at all frequencies).

However, this is nothing like what your ears would tell you. If the loudspeaker was just audible at 20 Hz, it would increase in loudness as the frequency rose, until at around 4 kHz it would sound (very roughly) 200 times louder. At higher frequencies still, it would get quieter again, finally fading into inaudibility at somewhere between 8 kHz and 20 kHz, depending how old you are and what you've been doing to your ears for the past few decades.

In practice, acousticians weight the response of the circuit of which the microphone forms a part, so that the system behaves like the ear—being most sensitive to frequencies at around 4 kHz. A frequency-weighted microphone is the heart of a sound level meter (SLM). There is actually a wide choice of different weightings, including some for dogs, but the most popular by far is A-weighting, which approximates the response of a human ear at moderate sound levels. Hence, the decibels that matter to us are usually A-weighted, written dBA, the full name for which is 'A-level weighted sound pressure level in decibels'.

SLMs are equipped with a choice of time integration factors. This matters because if a sound lasts less than about 0.1 second it sounds quieter, since the hearing system adds up the energy for about this period.

To add to the complexity, the loudness of a sound also depends on the nature of its source. For instance, people so dislike the sound of planes that, on average, they consider them to be as annoying as anonymous sounds that are about 5 dB louder. Conversely, people are rather fond of train noise, to the extent that they only find it as annoying as anonymous sounds which are about 5 dB quieter. These reactions are so well established that many planning applications which involve aircraft or railway noise adjust their figures by 5 dB (the corrections are known as the *aircraft malus* and the *railway bonus*). This means that no meter can actually measure what architects, home owners, noise campaigners, noisy machine buyers, and acousticians really need to know: how loud a sound is.

Considering all of this, there is little point in measuring SPLs with great accuracy: most SLMs are accurate to +/–1.4 dB at 10 kHz (called Class 2 meters). Even in laboratory work, +/–1.1 dB at 10 kHz is almost always enough (as provided by a Class 1 SLM). Far more important than accuracy is adherence to standard measurement procedures, including the frequent calibration of SLMs by comparison with standard measurement microphones.

Despite the complexities of loudness and its variation according to the source and the user, extensive surveys of the reactions of large numbers of individuals to carefully chosen sounds have determined roughly how loudness relates to SPL, and units have been defined on this basis, in particular the phon. Phons are defined as having the same values as the SPLs of 1 kHz tones, so a 1 kHz tone with an SPL of 10 dB has a loudness level of ten phons. But a 50 Hz tone with that same loudness level of ten phons has an SPL of 73 dB, because our ears are so much less sensitive at 50 Hz than 1 kHz that a 50 Hz tone needs to be 63 dB higher than a 1 kHz tone to sound as loud.

Loudness is just one of a large set of psychoacoustic measures, also known as sound quality parameters ('quality' being used in the sense of 'character' rather than 'goodness'). Loudness is by far the most commonly used and best developed; the others include sharpness (in acums), roughness (aspers), fluctuation (vacils), and dieselness (which has no units: different automobiles are simply ranked subjectively according to how 'dieselly' people think they sound). As the last-named suggests, these measures were developed primarily by the automotive industry in its attempts to make door-clunks, engine sounds, and even indicator noises sound appropriately powerful, masculine, reliable, and so on. In principle it would be very useful if domestic products and other noise sources could be characterized by such parameters.

The topic of sound qualities is part of the discipline of psychoacoustics, the study of the psychological effects of sound, which itself can be considered as an element of what is now known as sound studies. Sound studies deal with how sounds of all kinds have been made and consumed throughout history and in different cultures. Work on such topics has been carried out since the 1940s, and has increased greatly since the early 1990s.

Standing waves

A recurring aim in the history of acoustics is to make sound visible. In the 1780s, Ernst Chladni studied the ways in which metal plates vibrate when they are made to 'sing' by being stroked with a violin bow. Fine powder sprinkled on the plates is deflected from areas where vibration is strong, and collects in those that are still. The powder-free, strongly vibrating points correspond to antinodes (like the peaks or troughs in Figure 1), and the stationary, powdery areas are the nodes, the points where there is no pressure change (where the line crosses the axis in Figure 1).

It was possible for Chladni to 'see' sound waves in this way only because they did not progress through space: they were stationary, or 'standing' waves. For a standing wave, Figure 1 represents only how the pressure of the wave changes with location, and not the way the pressure at a particular spot changes over time (such a time-based diagram for any point in a standing wave would be a horizontal line).

The principle is clearer when considering the standing sound waves that are formed when one blows across the open end of a 12 cm tube which is closed at the other end. In all such waves, the air at the closed end of the tube cannot move, due to friction with the end wall (so this point is a node). The simplest wave of this kind is one in which the motion of the air molecules increases with distance from this end, and reaches a maximum (an antinode) at the open end. In this wave, one-quarter of a wavelength is inside the tube, so it has a wavelength of $4 \times 12 = 48$ cm. If one blows hard enough, a whole range of other standing waves will form, each with a node at one end of the tube and an antinode at the other, as shown in Figure 10. The lengths of these other waves are simple multiples of the first, and such waves are known as harmonics.

Just as in such a pipe, so in any other fluid-filled cavity, or any rigid object, there are certain wavelengths of sound which are

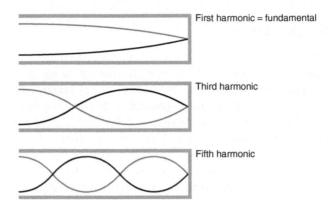

First harmonic = fundamental

Third harmonic

Fifth harmonic

10. Standing waves in a pipe open at one end.

particularly easy to excite. These are called resonance modes (or simply resonances), and the main ones can be predicted since they depend only on dimensions. For instance, if a 12 cm rod is fixed at its ends and struck sharply, it will generate 24 cm sound waves, together with waves of lengths 12 cm, 8 cm, 6 cm, 4 cm, and all others which include nodes 24 cm apart—again, a set of harmonics.

A 12 cm long box of air or water will produce all these waves too—in this case, what stops the 'ends' of the fluid moving is that they are adjacent to the box walls, where friction prevents free motion. The box will also produce families of waves corresponding to its height, width, and diagonals.

Resonances can be a major problem in room acoustics, but are the basis of most musical instruments. With an instrument that includes tubes open at one end (like some organ pipes), the open end is an antinode, and hence the fundamental frequency has twice the wavelength of that of a closed tube of the same length. (Actually, the antinode forms just beyond the pipe's end, requiring an *end correction* to be made, see Box 7.)

Box 7

End correction: $\frac{\lambda}{4} = l + c$ (closed pipe), $\frac{\lambda}{2} = l + 2c$ (open pipe).
End correction c, about 0.6 of the pipe radius.

Usually, the lowest resonant frequency is the most powerful; however, if a great deal of energy is supplied to an instrument, it may resonate an octave—or even two—higher. A flute, for example, will do this if it is blown hard enough ('overblown').

Resonances are all around us—tap a plate, glass, or fork and it will ring, provided only that it is not damped by being held too tightly (tuning forks still resonate if held tightly because they have two identical prongs that move in opposite directions, cancelling each other out at the handle so there is no resultant motion there). This is a handy way of finding whether crockery is cracked: if all is well, each successive millimetre of the plate will move immediately after its neighbouring millimetre, allowing waves to pass, much like a Mexican wave: the plate is, literally, sound. But if even a very fine crack separates adjacent areas, dragging and friction damp the resonance, giving rise to an unhealthy 'clink'.

If the force supplied to an object is a sound at the resonant frequency, the coupling will be highly efficient, hence guitar strings that sound in sympathy with those struck across the room, or bits of television sets that buzz annoyingly along with dramatic programme sounds.

An effect which is of importance in several areas of acoustics is Helmholtz resonance, familiar to anyone who has blown across the top of a bottle to make it sing. Any hollow object or cavity with an open neck will act as a Helmholtz resonator (Box 8). If a stream of air is blown across the opening, some will enter the neck, increasing the pressure in the cavity a little. This overpressure pushes the air out again—and, just like a pendulum,

Box 8

Helmholtz resonator frequency: $f \approx \dfrac{v}{2\pi}\sqrt{\dfrac{S}{lV}}$; sound

velocity v, neck area S, neck length (including end correction) l, cavity volume V.

this air 'overshoots' a bit, leaving a slight underpressure, which sucks more air in, and so on. This regular cycling constitutes a sound wave at a resonant frequency. If a sound wave at this frequency is supplied to the resonator, it will sound very strongly.

Charting sounds

Standing waves are a small subset of sound waves: mostly, the regions of high and low pressure in a wave move through space (such waves are called progressive or travelling waves). If one wishes to 'see' a travelling wave, one must therefore chart air pressure changes through time. One of the first to attempt this was Alexander Graham Bell, who in 1874 procured an ear from a corpse, impregnated it with oil to keep it flexible, and attached a thin straw to its drum. The other end of the straw was allowed to trace a line on a strip of soot-covered glass which was moved along as the ear was shouted at. This wobbly line was the first recording of a sound wave and the device was called an ear phonautograph. To the relief of those who had to construct them, later versions dispensed with dead ears in favour of metal diaphragms.

Phonautographs were no good for making actual measurements of sound waves, however; these were eventually provided by the cathode ray oscilloscope (CRO), developed in the 1930s. CROs can be set with different time-bases so that a high-frequency sound can be spread out across the screen or a low frequency one

11. Spectrogram.

compressed, so that their wave shapes can be seen. From this, their wavelengths can be read off and their frequencies determined.

Today, computerized versions of CROs are widely used. However, a two-dimensional plot can still only display some features of sound. Most sound waves vary rapidly both in frequency content and in pressure, which can only be properly displayed together on a three-dimensional display, called a spectrogram, which cannot be produced without a computer. In a spectrogram, height up the screen usually represents frequency, and brightness or colour represents sound pressure (or intensity). In other cases, a representation of a three-dimensional shape may be made on a screen, the results often resembling mountain ranges (Figure 11).

Unweaving sounds

Being able to see a sound allows one to find out a lot about it qualitatively, and rough measurements can also be made of the screen outputs, but often good quantitative information about sound is needed (perhaps to eliminate noise or improve the design of a musical instrument). For this, a mathematical analysis is required, and the most widely used and fundamental is based on work conducted by Joseph Fourier in the 1800s.

Fourier realized that any periodic function (that is, one which repeats at a steady rate) can be constructed by adding together a series (now called a Fourier series) of sine waves—and he worked out a method to determine what the members (*terms*) of that series are. (Mathematically speaking a Fourier series is composed of a series of sines and cosines—but a cosine is simply a sine wave which starts at maximum, rather than at zero, so I'll just refer to sine waves here.) As Figure 12 shows, as few as three sine waves can roughly approximate a square wave. To make the sides of the latter more vertical, higher-frequency tones must be added. A square wave sounds like a click, and Fourier analysis shows therefore that a sudden (that is, rapidly increasing in level) click will include some very high-frequency components.

Fourier's original work was only applicable to periodic waves, but a development of it known as the Fourier transform can handle non-periodic ones. A highly efficient mathematical method of calculating the component sine waves of a signal is

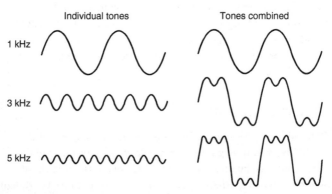

12. Summing sine waves to approximate a square wave.

known as a Fast Fourier Transform (FFT). When adding waves like this, their phase must be taken into account. During a single wavelength the sound pressure of a wave rises from zero (that is, equal to the ambient air pressure) to maximum, then falls down to minimum, and then rises to zero again. This is similar to the vertical motion of a dot painted on the edge of a rotating wheel, so phase can be described in terms of angles: starting at 0° the wave rises to a maximum at a phase of 90°, falls to zero at 180°, down to a minimum at 270°, and back to zero at 360° (which is the same as 0°).

All real sounds change over time, so the conversion into sine waves must be repeated frequently. Such time-varying frequency analyses of sounds have many applications: for instance, some of the parameters of the sound waves which compose an individual's voice are unique to that person. Hence, such parameters can be used as acoustic 'fingerprints' (called voice prints), and automatically recognized by a machine.

Conversely, since every word has a unique sound (except for homophones like 'sew' and 'so'), it is in principle possible for a machine to recognize them automatically, whoever speaks them. While the same word will be said differently by different speakers there are certain elements which vary only slightly, or predictably—hence our ability to recognize a word irrespective (within limits) of its speaker.

Automatic speech recognition is a long way from perfection however, and the main problem lies in deciding where one word ends and the next begins. To see how tricky this is, try listening to yourself saying 'bread and butter'. You will probably actually hear something like 'brembudder' with no silences at all (and saying the phrase 'properly' neither feels nor sounds natural). The reason that humans can identify words so readily is that the sound patterns we hear are only one piece of evidence as to what is being said—as Chapter 4 will explain.

Sounds from nowhere

Since any sound can be analysed into sine waves, it follows that any sound can be synthesized from them: synthesizers that generate speech from sounds have been available for many years, and work far better than recognizers. In practice though, it is often far easier to generate speech by adding together fragments of pre-recorded or pre-generated sounds—a technique known as voice coding.

Today's electronic systems can synthesize practically any sound at all, whether it occurs in nature or not—like the weird Shepard tone, which is produced by combining tones which fall in pitch but then fade out, while other, higher tones, fade in and themselves begin to fall. The impression is of a sound which continually falls—and yet gets no lower.

Usually, however, one does not want new sounds but improved versions of existing ones—a musical performance shorn of noise, for example. The selection of pre-recorded elements is also commonly used for non-speech sounds: one of the most celebrated electronic gadgets for the budding pop music producer in the 1960s was the *Mellotron* (1963)—a machine loaded with a library of short pieces of sound recorded on to magnetic tape, any of which could be quickly selected and played at a chosen frequency and volume.

Selecting sounds: filters

The commonest and easiest way to modify sound is through filtering: the removal or reduction of selected frequency ranges, accomplished either by electronic circuits or by software. High-pass filters cut out low frequencies, low-pass ones deal with high frequencies and band-pass filters banish both. A once familiar kind of variable filter is the graphic equalizer, a series of about seven slider bars on a hi-fi's amplifier, which allows selected preset frequency ranges to be suppressed. The simpler 'tone' control similarly quietens either high ('treble') or low ('bass') frequencies.

A vast range of other facilities is also to be found in the computerized toolbox of the sound artist or engineer: such software can add reverberation or echo to a recording, create an artificial soundscape, or apply such real-time changes as shifting the frequencies of the sounds of a pop song recording before passing to a loudspeaker. This is the basis of a karaoke system, in which the notes of songs can be sharpened or flattened to match those that the user finds easiest to sing.

Chapter 3
Sounds in harmony

What makes a note?

The words 'tone' and 'note' reflect the subjective/objective nature of sound: a tone is a sound wave with a particular frequency, a note is its subjective impact, with a particular pitch.

In addition to pitch, a note also has duration, loudness, and timbre. Timbre is by far the most important in identifying instruments by their sound, and is also the main carrier of emotional content. It is actually an uneasy bedfellow with the others, being far more complicated. It is a bit like defining a person by gender, height, weight, and fingerprint: the first three parameters can each be specified in terms of one value of a single unit, but the full description of an individual's fingerprint would be highly complicated and multidimensional. But, like a fingerprint, timbre is the only thing unique to a particular instrument or person, so it's the only thing that the brain *can* use as an identifier.

A problem for the brain which the fingerprint detective need not worry about is that timbre is a dynamic quality; it changes over time. The timbres of a cymbal, piano note, or drum all change radically from start to finish (this change is known as flux). The timbre of a particular instrument or speaker also changes with pitch and loudness: a cello's high notes have a different quality to its low ones, a shouted word is not simply louder than a spoken one, and a man's voice at the top of his range sounds much thinner than when he is singing low notes. Fortunately, we can rely on our highly evolved ability to focus on the very first sounds made by a source to untangle all this.

Prehistoric agendas

Our modern appreciation of music and our dislike of noise probably stem from the evolutionary pressures that moulded the hearing systems of our distant ancestors. Their first priority on hearing a sound would have been to identify it, and there was no time to ponder. The twang of a bowstring, the rumble of an avalanche, the thunder of approaching hooves or the warning hiss of a snake are only of benefit to the hearer if (s)he can react fast, and so the hearing system concentrates its analytical prowess on the earliest moments of a sound. This has the surprising consequence that the 'attack' sounds that an instrument makes in the first fraction of a second, though generally discordant and nothing like the instrument's 'steady state' sounds, are the ones that enable us to identify what that instrument is. A recorded piece of music edited to remove all the attacks sounds very strange, and the instruments could be anything. This fact stymied for many years attempts to synthesize instruments convincingly.

Often not just the instrument but the composition too can be identified within the first second. There was once a fairly popular radio (and then TV) quiz show called *Name that Tune*, in which contestants frequently succeeded in identifying tunes from just four or even three notes. It is very easy to outdo this if one is presented with a familiar recording: even a preliminary fumble or intake of breath is enough to identify it. Conversely, some musical sounds immediately inform the listener of the flavour of the whole piece: a single chord may have a 'Mexican' sound, a few violin strokes might sound 'folksy', a bagpipe skirl 'Scots', or a couple of notes 'fierce shark approaching'.

The responses to such brief cues are primarily emotional, as befits their original role as hazard warnings: the roar of a lion or the click of a safety catch being released deliver an immediate visceral response, releasing adrenalin in preparation for flight—or fight. In fact, our reflex response to *any* sudden loud sound close at hand is to put distance between us and it: a feedback shriek from a loudspeaker will literally make us jump away. Even continuous sounds are intrinsically repellent if they are loud enough: hence the almost physical reluctance one feels to pass close by a pneumatic drill.

The flipside to our focus on the very first parts of a sound is disinterest concerning sounds that last for a while, leading to the seeming paradox that a continuous sound at constant volume will sound quieter after a few dozen seconds: in nature, such sounds are likely to be as harmless as the breeze. But if that long-duration sound should suddenly stop, the silence is just as attention-getting as was its onset.

The act of listening to music has received considerable attention from a sound studies perspective. The musicologist and philosopher Peter Szendy, for example, argues that an essential part of listening to a piece of music is comparison with other works, performers, instruments—and other listeners too. His view is that, since the act of listening constantly 'appropriates' other things in this way, the essence of a piece of music can never be fully grasped.

Learning to sing

Musical sounds are far more important to many of us than those of other types—so much so that many of us can hardly restrain ourselves from bursting into song at the slightest provocation. Perhaps you have entertained yourself and, to a rather lesser extent, everyone within earshot, by singing 'Somewhere Over the Rainbow'. It's a fair bet that Judy Garland did it better: but how? Although your listeners may not enjoy the tune, they will surely recognize it because, even if a lot of the notes are wrong, they will at least go up and down in pitch in the right order, and vary in length in a familiar way.

Durations of notes (and of rests; gaps between notes) are indicated by their shapes, and defined as fractions of a 'whole note', which is a semibreve, as shown in Table 2.

There is no standard duration for a note, though in some cases a composer will provide one on an individual score: '♩ = 66', for example, means 'a crotchet lasts 1/66 of a minute'. Usually however, the only indication of the speed (technically, *tempo*) at which a piece should be played is a phrase (often in Italian) of variable helpfulness, such as *andante* ('at a walking pace') or *allegro non ma troppo* ('fast but not *too* fast'). Perhaps because the concept of tempo is not very applicable to natural sounds, we are poor at judging it—amateur or professional alike, spotting a 4 per cent difference in tempo is about the best we can do.

Unless you are a trained singer, or a naturally good one, you are probably singing some notes slightly 'flat' (that is, a bit lower than they should be) or 'sharp' (a bit higher). But what does 'should' mean here? It *doesn't* mean 'hitting the right notes'—it doesn't matter whether the frequency of the sound you make while singing '*Some…*' is 300 Hz or 333 Hz. But you do need to get the ratios of the frequencies correct. So, if you sing '*Some…*' at 300 Hz then you must sing '*…where*' at 600 Hz—that is, one octave higher—to avoid people wincing.

It follows from this that if you accompanied yourself on a piano without altering the pitches of your singing, you would almost

Table 2. Notes and rests

Sign	Name	Relative length	Rest
o	Semibreve	Whole note	▬
♩	Minim	Half note	▬
♩	Crotchet	Quarter note	𝄽
♪	Quaver	Eighth note	𝄾
♪	Semiquaver	Sixteenth note	𝄿

certainly be out of tune with the instrument. As far as a pianist is concerned, 'Some' is an F. There are eight F keys on an eighty-eight-key piano, and any of them may be used here—if the fourth F is chosen, the piano will play a note of 349 Hz.

If you do find yourself to be 'automatically' in tune with a piano, you probably have that rare thing: absolute (or 'perfect') pitch. Generating the same notes as your piano is all there is to having absolute pitch. If you were transported back two centuries, when pianos were not tuned to the same frequencies as they are now, you would of course not be in tune with them. And if you attempted to sing an unaccompanied duet with a singer of the period who had perfect pitch, the result would not sound pleasant.

Your pitch is nevertheless better than your 19th-century friend's in a certain sense: if you travelled two centuries into the future, your perfect pitch would match everyone else's, but your friend's would not. That is because, in 1939, at a London meeting of the International Standards Organization, many countries agreed on particular frequencies for particular notes. Like many other international agreements, this one only happened because the situation had become intolerable some considerable time before.

Since instrument makers, musicians, and composers had developed largely independently in different countries, each had adopted their pitches arbitrarily. In countries with several musical centres, pitches even varied from city to city. In 1780, a Berliner with perfect pitch would fall sadly flat in Vienna; a trumpet made in one city would be out of tune in the other—and trumpets cannot be retuned. Hence, as national and international travel among musicians became common, and successful instrument makers expanded their customer bases further afield, the need for an international standard became rapidly more apparent—hence the meeting, even on the eve of war.

It follows that absolute pitch cannot be an inborn talent; it must be learned. And learned before the age of six, usually by memorizing all the notes on a musical instrument. This is, on the face of it, an almost incredible feat of memory, and such memorization skills are alien to the adult mind. It seems that some children are predisposed to learn absolute pitch—but on the other hand there is a higher fraction of pitch-perfect persons in China and Vietnam than in Europe or the US, which may be due to the greater dependence in Eastern languages on the pitch at which words are pronounced.

For those of us who lack absolute pitch, it is the differences (*intervals*) between notes, not the notes themselves, that we memorize; hence a familiar tune is instantly recognized whether it is played on the double bass or the violin.

Why do we like what we do?

By far the most important interval in music is the octave, the interval in which one note has twice the frequency of the other. The primacy of the octave is demonstrated by the fact that, in the A to G notation system most of us use today, notes an octave apart are given the same letter. To distinguish different versions of each note, numerical subscripts are used as required (so the fourth F on the piano is F4).

Notes one octave apart are in harmony, and the impression we receive of a harmonious interval is that of auditory pleasantness: consonance. Other than a pair of identical notes, the octave is the most harmonious and consonant interval of all.

Exactly why harmonious intervals sound consonant is not quite clear, though we do know that the brain responds to harmonic sounds with synchronous neural firings. That is, the neurons which respond to each component of such sounds tend to synchronize their firing rates with each other. So, harmonious intervals are clearly important to us—but why do we find them pleasant, and inharmonious (dissonant) intervals to be less so? Actually, though we don't usually enjoy discords in isolation, the appetites of composers and their audiences for dissonance have grown steadily since the ancient Greeks—in fact, one could look at the history of Western music as the increasing adoption of dissonance (which is why listening to Stravinsky's *Rite of Spring* or Mozart's *Dissonance* quartet is not as disturbing as it once was). Nevertheless, if one ranks intervals by their perceived consonance, the result is pretty independent of geography or history. So there is there is *something* objective about consonance. In fact, two things:

1. A lack of 'roughness'. When the ear receives two tones close together in frequency (about 5 per cent different, or two or three semitones), they are perceived as a single rough-sounding note.
2. Similarity to a harmonic series, which is a set of tones with frequencies such as 1 kHz, 2 kHz, 3 kHz, 4 kHz, etc.

More notes

Music making cannot rely on octaves alone. For one thing, we can't hear more than about ten of them. Also, producing such a range is a challenge: violins and guitars cover about four octaves, exceptional humans can manage five or six, pianos seven or eight. Furthermore, every tune would be nothing but the purest harmonies—which very soon become boring.

Once the octave had been defined, Pythagoras and others set about adding notes to it, embarking on a long research project that was to last for well over a thousand years. Although Pythagoras actually used a monochord, it's simplest to the job with several strings of variable lengths, all with the same tension and width and all made of the same material.

The next most harmonious/consonant pair of notes after the octave is that given by a pair of strings in which one string is 2/3 the length of the other. Three other simple ratios of string lengths yield other fairly nice-sounding note combinations, and the results are shown in Table 3.

Table 3. Harmonious intervals

Name of interval	Length of shorter string compared to first string	Frequency of note from shorter string compared to note from first	Order of consonance
Unison	1	1	1
Second	8/9	1 1/8	6
Third	4/5	1 1/4	5
Fifth	2/3	1 1/2	3
Fourth	3/5	1 2/3	4
Octave	1/2	2	2

If we were to build a harp with strings of lengths defined in Table 3, but all of the same thickness, tension, and material, we would very easily be able to make simple, reasonably pleasant-sounding chords with it—in fact, any pair or larger group of such strings played together wouldn't sound too bad.

This system is called the pentatonic (five-tone) scale. It was discovered independently by numerous ancient musicians in many parts of the world and is still very popular today ('Amazing Grace', 'My Girl', and 'I Shot the Sheriff' are written in it). The black keys on the piano are a series of pentatonic scales, so if you press them randomly it won't sound as horrible as if you do the same with the white ones.

Even more notes

Although people like musical harmony, they also like a bit of discord, and the pentatonic scale was just too tuneful and safe for composers who wanted to challenge their audiences and push the boundaries. Also, five notes per octave is still not that many: but how to add more? Answering this question initiated centuries of careful experiment and academic debate, leading finally to a system called equal temperament. In this system, an octave is divided up into twelve notes, each a semitone, or one hundred cents, apart.

In keeping with the idea that what matters in music are ratios of notes, not notes themselves, semitones and cents are defined in terms of ratios. Each successive semitone is about 6 per cent higher than the one before. This means that the higher the number of hertz corresponding to a semitone, the higher the note is.

A few composers have experimented with more than twelve notes per octave, and extra notes may appear briefly, in moving gradually from one note to another to another (this is called a glissando, or slide). However, in all cultures, the number of 'proper' notes per octave is almost always twelve.

Having defined twelve notes, people proceeded to use just eight of them per octave (hence the name). There are only seven *different* notes in an octave—the eighth is the same as the first, just one octave higher (even professional musicians who can immediately identify a note struggle to say which octave it is).

The reason why we use only eight of the twelve notes in an octave was resolved in 1956 by George Miller, a psychologist, who found experimentally that our short-term memory can store no more than about seven items—hence seven different notes—at a time.

There is a semitone jump between adjacent notes in the full group of twelve, so any group of seven notes that we choose to make our scale will include some which are two semitones (one tone) apart. So, we might have this scale: A *gap* B C *gap* D *gap* E F *gap* G. (A scale is any sequence of notes in which each is higher than the preceding one.)

The fact that we have chosen a set of seven different notes to define our scale does not mean that we may not use other notes in a piece of music. If we want to use those 'gap' notes, we mark them with symbols which mean 'a semitone higher than the next lowest note in the scale' or 'a semitone lower than the next highest note'.

The optional extra note that fills the first gap in our scale above is called either A sharp (written ♯) or B flat (written ♭). Notes may be further subdivided if required; such smaller fractions are referred to as microtones.

In writing a piece of music there is usually a note that matters more than the others—the key note, or tonic. Usually, this note is the one on which a song begins and ends. In 'Somewhere Over the Rainbow', the key note is the one sung on 'Some' and 'I'. The way Judy Garland sings it in the film, this is E flat (E♭). Hence, her version is said to be in the key of E flat.

How is a scale chosen? The main consideration is the mood of the piece: if the composer is writing a confident, celebratory, up-beat piece, (s)he is likely to choose a major scale. To write a piece which is less emotionally clear a minor key is usually chosen. The primary objective difference between major and minor scales is that a (natural) major scale consists of a tonic (which can be any note), plus the six notes which 'fit best' with it, in the sense that their wavelengths bear the maximum number of the simplest possible whole-number ratios with each other. The sound waves hence combine to give the most regular/even patterns possible. In practice, one takes the pentatonic and adds two more notes. One of these two notes, the leading note, has a frequency one and seven-eighths that of the lowest note.

In a natural minor scale, two of the notes of the pentatonic, plus the leading note, have been lost (they are down-shifted by a semitone). The result is that there is no longer a big set of simple whole-number ratios, and the impact is vaguer, more complex, and less complete-sounding.

Additionally, when we hear any sung or played note, it is invariably (except in the case of some electronic instruments) accompanied by its full set of harmonies. Major scales contain all such harmonies too, but minor scales do not. Hence, there is a genuine 'naturalness' or 'completeness' in major scales which is lacking in minor ones. 'Amazing Grace' and 'My Girl' are both in major scales, 'I Shot the Sheriff' is minor.

It used to be (and occasionally still is) claimed that individual keys have their own emotional qualities (in addition to the distinction between major and minor). Numerous experiments have demonstrated that this is not so. It is likely that this idea arose through the reputations of famous or influential pieces of music—imitations tended to be in the same key, which hence became associated with the emotional quality of the original piece. That is not to say that the choice of key is arbitrary: some instruments are designed so that the notes of a particular key are physically easier to play.

Sound sequences

While the raw materials of musical composition are similar the world over, there has been a great difference in its development: in Western music, relatively few notes are used, but they are often played together (chords) or in relation to each other (harmonies). If we had to juggle more than seven distinct notes, musical composition would be more difficult. in particular, polyphony—the simultaneous playing or singing of different melodies—would become well-nigh impossible. In Eastern music, many more notes tend to be used, but combinations are rarer. This difference in approach fed through to instrument design: Western instruments like the guitar and organ are naturally polyphonic, while Eastern ones, like the sitar, can often only produce one note at a time, but can easily vary that note rapidly and subtly.

The differences between Western and other music may have their origins in the 10th century, when polyphony began in Europe (though there are some who claim that polyphony has far earlier roots). From polyphony sprang both the characteristically multi-tune approach of the West and the studies of harmony required to develop it. It has been suggested that the reason for the first experiments with polyphony was a desire to have men, women, and children singing together.

Melodies

Play some notes in sequence, and you have a tune—technically a melody. If we forget about the different pitches, then what is loosely called the rhythm is all that remains—the relative lengths of the notes and rests, the speed (*tempo*) with which the whole thing is played, and the pattern of beats (*metre*, or sometimes *measure*). This is not quite the technical meaning of the word rhythm, which refers only to the length pattern. In 'Somewhere Over the Rainbow', the rhythm (technical version) is: 'long, long, medium, short, short, medium, medium'.

The metre is indicated by the time signature. The time signature of the commonest metre ('common time') is written $\frac{4}{4}$. The lower number gives the units used to specify the beat: 1/4 notes (crotchets). The upper specifies the number of these beats per group. In a score, the end of each group being shown by a vertical line called a bar (so common time is 'four crotchet beats to the bar').

These are not the only kinds of beats that matter in music—the other kind occurs when two tones of similar wavelength play together. Every so often the peaks or troughs of the two sound waves will coincide, adding together to make regions of extra-high or -low pressure. The effect is heard as a low-frequency 'beat', the frequency of which is the difference between the frequencies that make it up (Figure 13).

Watch any black and white thriller featuring London policemen, and you'll hear an example of beats when one of them blows his whistle: such whistles used two pipes of slightly different lengths to produce their arresting sound. But beats aren't always unpleasant: some organs include a stop called the *voix céleste*,

Sound

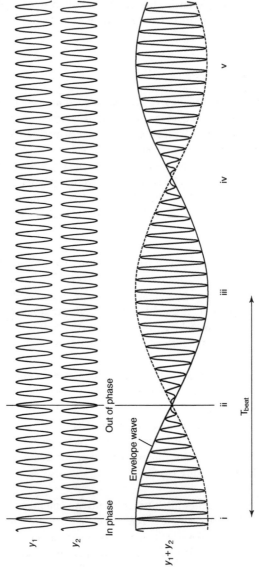

y_1

y_2

In phase   Out of phase

Envelope wave

$y_1 + y_2$

i   ii   iii   iv   v

T_{beat}

13. Beats.

which controls a set of slightly out-of-tune pipes. When played along with normal pipes, the beats produced make a pleasantly wavering sound.

Instrumental break

The conventional way to classify musical instruments is partly by what they were originally made of, hence strings, brass, woodwind, and miscellaneous (percussion). Another way is by whether an instrument is pitched (like a piano) or unpitched (like maracas). Any piece of music needs rhythm, and using a non-pitched source avoids dissonances with the 'melody' instruments. It is not essential that the rhythm source be a musical instrument: Jean-Baptiste Lully (1632–87) used to thump a big stick on the ground. This is no longer fashionable, perhaps partly because Lully died from it, having hit himself on the toe and contracted terminal gangrene as a result. Drums soon became a more popular source, though they are also used as a source of the 'actual' music. For this reason, while drums such as bass drums and snare drums are unpitched, kettle drums are pitched.

Most pitched instruments make their notes either by striking, plucking, or bowing a string, wire, plate, or other solid object, or making a column of air vibrate by providing it with energy from lungs or bellows. In every case, it is resonance plus overtones (and, in the case of bells, undertones, also called subharmonics) that provide the required sound.

This makes it seem simple to construct a musical instrument, but there are three complicating factors. The first is that an object such as a metal plate will not only resonate at a wavelength twice its length, but also at wavelengths twice its width and thickness—and it will produce other sounds as it twists, too.

The second point—a very useful one for string players—is that when an object such as a 120 cm string is struck (or plucked or bowed) it will move in a pattern determined not only by its length but also by the position of the striking point, since that point has no option but to be an antinode. The ends of the string have no choice but to be nodes, so if the string is struck at a point 40 cm from an end, it will make a fundamental with a node-antinode distance of 40 cm, and hence a wavelength of 160 cm. There will be another sound with a node-antinode distance of 80 cm. These components will fade rapidly as the string's vibration shifts towards its natural mode of vibration, which is to have an antinode at the centre and hence make a 240 cm wavelength tone. The subjective effect is that near-centre plucking produces a mellower sound, near-end plucking a harsher one.

Pipes and tubes behave in a somewhat similar way. In an open pipe, the simplest motion is one in which there is free movement at the open end and none (since there can be none) at the closed end: so a 12 cm pipe will have a fundamental with a node at 0 cm and the first antinode at 12 cm (plus a bit due to the end correction), hence a wavelength of about 48 cm. But, just as in a string, this is not the end of the story: though the fundamental is defined by the pipe's length, other resonances will be produced according to its width. The mixing of these with the fundamental and its overtones produce different timbres: in wider tubes, there is more energy in lower harmonics and the timbre is more rounded, while thinner pipes are brighter (or shriller, depending on your taste).

The final consideration for instrument makers is that a string held between two points makes a very quiet sound indeed, because its very thin cross section only affects a tiny surrounding volume of air. In order to turn the string into a musical instrument, a much greater volume of air must be put in motion. In an electric guitar, this is accomplished by using the vibration of the metal string to induce a small electric current in pickups (magnets wound with fine wire). In acoustic guitars and other stringed instruments, the wooden soundboard (top) is vibrated by the string and its large area moves much more air than the string alone. Also, at low frequencies the hollow bodies of such instruments act as Helmholtz resonators.

Non-electric instruments do not add energy. Rather, they convert the energy provided by bow, plectrum, or air blast to an acoustic form. In cases where the body of the instrument shifts the frequency towards the 4 kHz region, this will sound louder, despite the fact that the sound power is actually reduced a little by frictional losses. This loudness increase stems from the construction of our hearing systems, as we shall see in Chapter 4.

Chapter 4
Hearing sound

The range of hearing

Being able to hear is unremarkable: powerful sounds shake the body and can be detected even by single-celled organisms. But being able to hear as well as we do is little short of miraculous: we can quite easily detect a sound which delivers a power of 10^{-15} watts to the eardrums, despite the fact that it moves them only a fraction of the width of a hydrogen atom.

Almost as impressive is the range of sound powers we can hear. The gap between the quietest audible sound level (the *threshold of hearing*, 0 dB) to the threshold of pain (around 130 dB) is huge: 130 dB is 10^{13}, which is the number of pence in a hundred billion pounds.

We can also hear a fairly wide range of frequencies; about ten octaves, a couple more than a piano keyboard. We can detect, though not truly hear, frequencies well below and above this too, as will be explained in Chapter 6. And our frequency discrimination is excellent: most of us can detect differences of about a quarter of a semitone; with practice and in ideal conditions, a difference of about one-twentieth of a semitone is just distinguishable. Our judgement of directionality, by

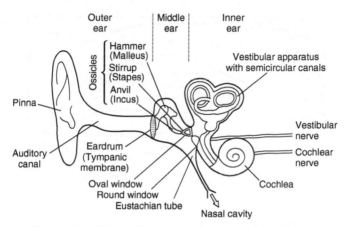

14. The ear (the middle and inner ears are greatly enlarged).

contrast, is mediocre; even in favourable conditions we can
only determine the direction of a sound's source within about
10° horizontally or 20° vertically; many other animals can do
very much better.

Perhaps the most impressive of all our hearing abilities is that we
can understand words whose levels are less than 10 per cent of
that of background noise level (if that background is a broad
spread of frequencies): this far surpasses any machine.

Our ears have two functions, hearing and balance, and balance is
taken care of entirely by the semicircular canals (see Figure 14).
The rest of the ear has thus been free to evolve the best possible
system for our needs. Not so our vocal apparatus; as a relative
latecomer in our evolution it had to fit itself in cheek by jowl,
sharing structures previously earmarked for breathing and eating,
licking and sucking, kissing and fighting. Yet, in a trained actor or
accomplished singer, the speech system functions just as perfectly
as a Stradivarius violin.

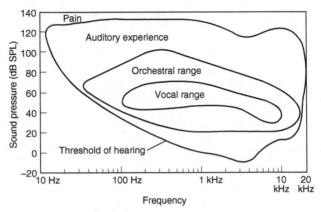

15. Frequency dependence of hearing.

What do ears hear?

As Figure 15 shows, our hearing systems have not evolved to measure the physical powers of sounds: a piccolo, for example, has a maximum power of about 0.08 watt, while a trombone can manage about 6 watts, yet it sounds the quieter instrument. This means that trombone players must work a lot harder than piccolo players (who in fact never play at full blast in any ordinary piece of music). But all instrumental powers are low. A loud orchestra playing at full stretch might manage 60 watts; if it could keep it up for two minutes, that would be enough to boil a tablespoon of water.

Moreover, this power spreads from the source in many directions, so, if you were sitting 10 metres from such an orchestra, less than 0.01 per cent would reach your eardrums. And what they actually detect is nothing more than a series of rapid prods of air, which carry no information other than how hard they hit and how rapidly they arrive. That we can experience a whole world of sound, charged with emotion and packed with meaning, is

54

due to the precise coordination of highly evolved anatomical, electrochemical, and neurological processing systems.

The outer ear: capturing the sounds

The pinna, the visible part of the ear, acts mainly as a funnel to collect sounds; its front-to-back asymmetry also provides some information about their direction. The moveable pinnae that some mammals possess greatly assist in direction finding; though some of us also have this ability too, it is of no benefit beyond its entertainment value.

Since the auditory canal is a cylinder around 30 mm long, it has a resonance at a wavelength twice this, corresponding to a frequency around 3 kHz. The gain in energy at this frequency results in a loss at others, so the canal acts as a partial band-pass filter.

At the end of the canal is the eardrum (*tympanic membrane*), a roughly circular disc of stretched skin about 1 cm across. It is inclined at an angle to the canal, to maximize its area and so capture as much force as possible. It is also slightly conical, which allows it to transfer more energy than it could if it were flat. The drum has no resonances—for frequencies above ~3 kHz, its surface moves in a chaotic way, and for lower frequencies it all moves as one piece, since the waves are larger than it is. As a result it transfers the widest possible range of frequencies with minimal filtering. This near-flat frequency response is achieved in part through its asymmetric shape and in part through an internal scaffold of collagen fibres.

It is important that the drum is taut but not stiff, and this is made possible by the equalization of internal and external pressure by the Eustachian tube (always drawn as open despite the fact that it never is, except for an instant when there is a significant pressure change—making a distinctive crackling sound when it operates).

The middle ear: sharpening the blow

The eardrum is connected to a series of three tiny (tiniest, in fact) bones called ossicles, which occupy the air-filled middle ear. Their main job is to covert the wide shallow motion of the eardrum to a higher-pressure 'tap' on a second drum-like membrane called the round window, which is the gateway to the inner ear. By working as levers, the ossicles increase the force by about 1.5 times. However, the main way in which the force is enhanced is simply through the ratio of the areas of the eardrum and the round window, which increases the pressure about twenty times by concentrating the force over a smaller area. The ossicles also provide some protection for the inner ear, through the acoustic reflex (see Chapter 8).

The inner ear: sound to electricity

The inner ear is full of liquid, and, just as the eardrum converts airborne sound to bone-borne, so the round window converts the latter to a fluid-borne version. This passes along the cochlea, a coiled tube about 2 cm long. At its extremity is a hole (the *helicotrema*), and the wave passes through this and then travels back along a second tube which is joined along its length to the first. When the wave has completed its doubled-back journey, it must be eliminated, otherwise it would reflect back up the cochlea and interfere with newly arrived waves. So, the second tube terminates in yet another membrane, the oval window. This bulges outwards when the waves reach it, dissipating them as heat.

Running between the tubes like the filling in a baguette is the basilar membrane, which converts the waves to nerve impulses.

The side-by-side tubes would be about 5 cm long, were they not coiled up like a snail (*cochlea* is Latin for snail). Because the length of the cochlea is related to the wavelength of sound waves, it is similar in all mammals: only about 50 per cent longer in an elephant than a person, so the coiling is probably simply a space-saving measure, rather than having any acoustic function. Mice and other tiny mammals cannot accommodate a full-size cochlea. Theirs are about 1 cm long—and consequently they can only hear about three or four octaves compared to the eight to ten octave range which we and most other large animals can detect. On the membrane is mounted the organ of Corti, on which grow around nine rows of short hairs called stereocilia (around 400 per row). These rows run the length of the membrane, and have nerve fibres attached to them. These fibres bundle together to form the auditory (cochlear) nerve, which transmits impulses to the brain.

The basilar membrane moves in response to the sound waves that impinge on it. It is stiffer and wider at its root than at its tip, which means that lower frequency sounds cause oscillations closer to the latter. These oscillations cause the stereocilia to move, and the hair cells to which the stereocilia are attached then send electrochemical impulses to the brain. Since the brain knows which cells are where, it can determine the frequency of sounds by this means (this is known as place theory). However, when presented with sounds below about 1 kHz, the whole membrane oscillates. In this case, a different mechanism becomes important, in which the hair cells fire in time with the pulses that make up the sound wave—firing one hundred times a second in response to a 100 Hz tone for instance.

However, the cells are incapable of firing more rapidly than about 500 times a second. To respond to higher frequencies they therefore face a similar problem to that of a squad of soldiers armed with flintlocks which take, say, a minute to reload: how can the squad produce sustained gunfire at intervals of 10 seconds? The answer is to divide the squad into six groups. The first group fires and starts to reload. Ten seconds later, the second group fires, and so on. Ten seconds after the sixth group has fired, the first group will have completed its reloading and will fire again. Hair cells work in groups in just this way: the first group might signal the brain when a cycle of the sound wave is at its peak, the second when that cycle has fallen halfway to its minimum, the third at minimum and the fourth when it is halfway back up to maximum. In this way, such a quartet of cell-groups could respond to a tone four times higher in frequency than their individual maximum firing rates.

The hair cells of the basilar membrane also work together in a very different way: those in the eight or so outer rows respond to incoming sound waves by changing their lengths in time with them. This motion amplifies the vibrations of the stereocilia on the inner row of hair cells (the only ones that send signals to the brain) and hence provides significant amplification, allowing us to hear sounds which are 40 dB (one ten-thousandth) less powerful than we otherwise could. This activity generates its own faint sounds, called otoacoustic emissions. These emissions are of great value in determining the functionality of the hearing system in infants who are too young to report whether or what they can hear. Also, they often fade when there is any damage to the inner ear, so they are a useful check for audiologists (hearing specialists) too. They are (fortunately) far too quiet to be heard, so sensitive in-ear microphones are used to measure them.

Nerves and brain: objective to subjective

The nerve signals that emerge from the basilar membrane are not mimics of sound waves, but coded messages which contain three pieces of information: (a) how many nerve fibres are signalling at once, (b) how far along the basilar membrane those fibres are, and (c) how long the interval is between bursts of fibre signals. The brain extracts loudness information from a combination of (a) and (c), and pitch information from (b) and (c).

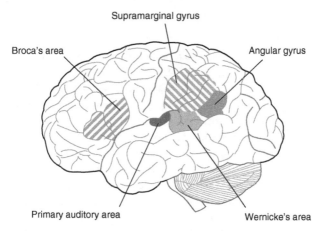

16. Locations of hearing, language, and speech activities in the brain. The primary auditory area recognizes sounds. Broca's area both analyses and produces semantic elements. Wernicke's area deals with the sequences of speech sounds (heard, made, and remembered). The supramarginal gyrus may deal with articulation, and the angular gyrus may assist with semantic processing.

What happens in the brain's hearing and language centres (see Figure 16) is not entirely clear, but the first stage in processing is to extract salient features from the stream of input data from the auditory nerve. These features are then used to continuously update, amend, and refine a mental model of the thing being listened to—perhaps a tune, a spoken phrase, or a worrying engine noise. The model's accuracy is tested by predicting what the sound will do next. What the brain is trying to do is establish the degree to which each component of a sound contributes to the meaning, a process called hierarchical encoding. Our incredible prowess in performing this function is demonstrated by our ability to, for example, recognize that someone is trying to hum 'Somewhere Over the Rainbow' even if they get most of it wrong. The speed with which the mammalian brain interprets the sounds that reach the ears is almost incredible—say 'squirrel' to a dog who knows the word, and the response is almost instantaneous.

Homing in

Following a piece of music is a rather unusual and newfangled (in evolutionary terms) thing for a brain to do: usually the focus of interest is an object in the external world, and consequently an important function of the hearing system is to locate that object. For high frequencies, the brain relies partly on the fact that a sound coming from the left will arrive first at the left ear, and partly on the blocking effect of the head, which means that the level at the right ear will be lower. For a sound wave longer than the distance between the ears, the brain compares the changing levels of sound at the ears: if a long wave arrives from the left, each of its antinodes will reach the left ear first, so the pressure at that ear will initially be highest. As the wave progresses, the left ear pressure falls while the right rises until the antinode passes it, when it will begin to fall again. However, with a wave more than about 4 metres long, there is very little change in level over the inter-ear distance, so its direction cannot be judged.

To determine whether the sound source is above or below the ears, the brain relies on the effects of the shapes of the head and shoulders on the level of the sound. Direction finding is not the only advantage of having two ears: the auditory nerves sometimes fire even when no sound is present, but the brain will reject such signals if they come only from one side.

The brain's processing system has evolved to make reasonable assumptions about the sounds it receives, leading to such phenomena as the precedence effect (also known as the law of the first wave front or the Haas effect). The brain's assumption is that a sound that arrives in the first fraction of a millisecond indicates the direction of the sound source. So, subsequent sounds are regarded as coming from the same direction as the first. This allows us to locate a sound source in dark spaces without being confused by echoes arriving from many directions. Such assumptions can mislead, especially in situations that would not occur naturally: for instance, if one listens to a sound from a loudspeaker about a metre away and at 45° to the left, and this sound is gradually replaced by an identical one from a loudspeaker at 45° to the right, the sound will still seem to come from the left.

A very useful feature for people trying to communicate in crowds is the cocktail party effect, in which particular phrases (such as one's name) stand out from the hubbub. This works for non-vocal sounds too: conductors are often highly attuned to particular instruments or musical phrases. This effect works because the brain is constantly model-building whether we are actively listening or not, and because it preferentially seeks matches with sounds that it has classified as having a significant meaning, like its owner's name.

The role of hearing is a very rich and complex one—as linguist Roland Barthes points out, sounds act on our minds in three ways: as 'indices' (the alarming sound of an explosion), 'signs' (the literal meaning of a word), and 'signifiers' (unconscious associations triggered by a word like 'end'). And hearing is also very much a social activity too—according to Labelle:

> the rich undulations of auditory material do much to unfix delineations between the private and the public. Sound operates by forming links, groupings and conjunctions that accentuate individual identity as a relational project…[and] weave an individual into a larger social fabric…contributing to the meaning of shared spaces.

Labelle points out, too, that whether we like it or not, sounds bring us into intimate contact with others: crying babies, noisy neighbours, or cheering fellow football supporters. As he puts it: 'Sound creates a relational geography that is most often emotional, contentious, fluid.'

Hearing bones

The eardrums are wonders of evolutionary engineering, but we can actually hear fairly well without them, since sound waves also reach the inner ear by travelling through the bones of the head, specifically the mastoid bone behind the ear. Submerging one's ears in the bath largely switches off the airborne route so the sounds that remain arrive mainly by bone conduction. This system is rather insensitive however: using air conduction, we can hear sounds about 40 dB less powerful than the weakest detectable through our mastoids. On the other hand, bone conduction allows us to detect sounds with frequencies up to 30 kHz, which is well beyond the upper frequency limit of the airborne route—but, presumably because such sounds are of little value to us, they are all encoded in the same way, so they give rise to the same pitch sensation as 20 kHz sounds.

Bottlenose dolphins (*Tursiops truncatus*) take bone-based hearing much further: their jaws bear teeth that are spaced at regular intervals and set at the same angle, are all of a very similar shape, and have heights that depend upon their location. This adds up to a *focussing array*, in which sound waves of particular wavelengths are significantly amplified—but only if the source is directly ahead. Thus, the dolphins can hear very quiet sounds, and can determine their directions simply by moving their heads until the loudness is maximized.

Deafness

The hearing system is a delicate one, and severe damage to the eardrums or ossicles is not uncommon. When it occurs we must rely instead on bone conduction and artificial aids: Edison used his teeth to transfer the sounds from his phonograph to his mastoid, as the marks on the device still testify.

This condition is called conductive hearing loss. If damage to the inner ear or auditory nerve occurs, the result is sensorineural or 'nerve' hearing loss. It mostly affects higher frequencies and quieter sounds; in mild forms, it gives rise to a condition called recruitment, in which there is a sudden jump in the 'hearability' of sounds. A person suffering from recruitment and exposed to a sound of gradually increasing level can at first detect nothing and then suddenly hears the sound, which seems particularly loud. Hence the 'there's no need to shout' protest in response to those who raise their voices just a little to make themselves heard on a second attempt.

Sensorineural hearing loss is the commonest type, and its commonest cause is physical damage inflicted on the hair cells. With very high levels of sound, the eardrums can be ruptured (and they can also be damaged by blows to the head or by infection). Remarkably however, burst eardrums can not only heal, they usually then perform almost as well as before.

About 360 million people worldwide (over 5 per cent of the global population) have 'disabling' hearing loss—that is, hearing loss greater than 40 dB in the better-hearing ear in adults and a hearing loss greater than 30 dB in the better-hearing ear in children (who make up about 10 per cent of the total). About one in three people over the age of sixty-five suffer from such hearing loss.

In terms of alleviating deafness, what can be done varies greatly. For conductive hearing loss, if some hearing function remains, hearing aids tailor-made to fit the wearer's particular pattern of loss are highly effective. They may also be equipped with noise-cancelling functions and often use directional microphones, so that the user can focus on sources that they are looking at. They can also include a vibrating element to stimulate the mastoid. Hearing aids are less successful in treating sensorineural hearing loss, because signals are distorted by recruitment.

With complete deafness the challenge is much greater, but in the last few years the introduction of fairly reliable cochlear implants has given new hope. And the future looks promising: in 2012, by encouraging stem cells to grow into hair cells, deafened gerbils had their hearing restored, typically by 45 per cent, but by 90 per cent in a few cases. This approach may one day be applicable to those people (around 10 per cent) whose hearing loss is caused by damage to nerves called spiral ganglion neurons. There are also some animals, including owls, in which lost stereocilia simply grow back, and it may be that this capability could be genetically induced in humans (without adopting the other peculiar feature of owl hearing, which is that it only works at its best in spring, presumably because owls need to catch extra prey for their chicks then).

There are a great many hearing problems which are not simply characterized by loss of function. The commonest is tinnitus: ringing in the ears. Its causes are largely mysterious, and the level, type, and duration of the sounds vary greatly from person to person. It is often associated with past infection, drugs (especially certain antibiotics), or trauma, and it frequently accompanies hearing loss.

Structures for speech

Around a million years ago, the hearing systems of our ancestors underwent subtle changes to fine-tune them—literally—to detect speech. We know little about the evolution of our vocal systems; although the ability to make sounds in a controlled way is common to most animal species, speech is immeasurably more complex. So, unlike the evolution of, say, the leg, we cannot look back at a long chain of ancestral forms and watch the system adapt to the changing requirements of those users and the shifting demands of their environments.

In its most basic form, the making of sounds is simple: Figure 17 shows the structures involved. Air exits from the lungs through a

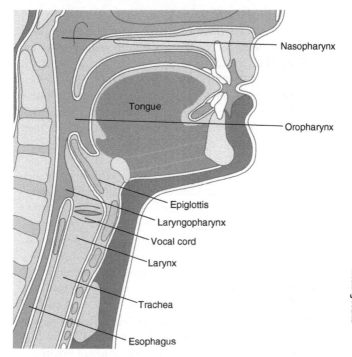

17. Vocal apparatus.

tube—the windpipe or trachea—which is equipped with flap-like
vocal folds (vocal cords) that restrict the air flow and vibrate when
they are tensed by muscles attached to them. Increasing this
tension increases the vibration frequency, but the length of the
folds sets its lower limit, resulting in fundamental frequencies
round 125 Hz for men, 200 Hz for women, and 300 Hz or above
for children. In boys, there is a sudden increase in length during
puberty, which causes the 'breaking' of the voice.

After emerging from the end of the trachea, this low-frequency
sound enters the back area of the vocal tract, roofed by the soft
palate. In front of this is the hard palate. Together these structures

form a cavity in which the sound forms resonances called formants. The characteristic wavelengths of vowels are set here, and varied by raising the tongue to change the volume of the tract or to divide it into two linked cavities.

Consonants involve more parts of the vocal apparatus than vowels, are usually shorter in duration, and, in many cases, change while they are being made. There are four main types, defined by their manner of articulation.

Plosives are made by the sudden stopping of airflow (hence their alternative name: *stops*). *Fricatives* and *liquids* require partial stopping, with or without turbulence respectively. *Nasals* deflect the airstream to the nasal cavity. *Glides* (*semivowels*) involve a rapid transition from one vowel sound to another. The unusual helpfulness of this naming system is shared by the subdivision of the consonants according to their place of articulation, as Table 4 shows. Also shown is whether or not the consonant is voiced—that is, whether the vocal cords are involved in making it. (Ventriloquists attempting to produce labial or labiodental sounds are stymied by the need to keep their lips separated and motionless. Skilled proponents of the art circumvent this by speaking very fast indeed at such points.)

Since the wavelengths of the resonances in the vocal tract depend only on its structure, changing the velocity of sound alters the frequencies of those wavelengths. Hence the 'Donald Duck' voice produced by breathing helium (14 per cent of the density of air), and the more rarely heard gravelly voice produced after breathing xenon, which is 4.6 times denser than air.

Life would be a dull thing, however, if all we did with our vocal apparatus was speak. *Singing* is, physiologically, no different from speaking except that every aspect of the sound made is more finely controlled, and pitch is often keyed to an externally defined value. *Whistling* does not involve the vocal cords: it requires the

Table 4. English consonants

Manner of articulation		Place of articulation						
		Labial (lips)	Labio dental (lips and teeth)	Dental (teeth)	Alveolar (gums)	Palatal (hard palate)	Velar (soft palate)	Glottal (glottis)
Stop	Voiced	b			d		g	
	Unvoiced	p			t		k	
Fricative	Voiced		v	th(en)	z	zh		
	Unvoiced		f	th(in)	s	sh		h
Nasal	Voiced	m			n		ng	
Glide	Voiced	w			y			
Liquid					l & r			

production of turbulence around the lips, which transfers energy to the vocal cavity, which acts as a Helmholtz resonator. *Shouting* simply requires greater air force from the lungs. In *whispering*, the vocal apparatus works as it does when producing normal speech, except that the vocal folds are neither vibrated nor fully relaxed, so that when air passes through them it produces turbulence (this is called adduction). Since much more air can pass between the folds without exciting sound waves, whispering is necessarily relatively quiet.

Our hearing systems are far more sophisticated than our most advanced machines, and have evolved to suit us admirably. But what nature has given us is limited in range. In all but the tiniest groups of people, communication—which we prize so highly—must spread far beyond the reach of voice or of hearing, and it was to answer that need that first electricity and then electronics were pressed into service, to augment and to replace our flesh and our nerves. How they do this is the subject of Chapter 5.

Chapter 5
Electronic sound

Sound to electricity: microphones

As invented by Charles Wheatstone in the 1820s, the microphone was a purely acoustical device made of two metal plates clamped to the ears by a springy rod and a length of dressmaker's ribbon. To use it, one pressed one's head against the sound source (Wheatstone helpfully suggested a boiling kettle) which, with a bit of luck, could be heard more clearly. To no one's surprise but Wheatstone's, it didn't catch on.

The ancestor of today's devices—which convert sounds to electricity rather than simply directing them to the ear as Wheatstone's version did—was the carbon microphone, invented by David Hughes in the 1870s. In this, a thin metal plate (the diaphragm) compresses a vessel filled with carbon granules through which a current is passed, and the compressive force alters the electrical resistance. Although their performance was very poor, carbon microphones were used in telephones for decades.

Although there have been many microphone designs over the years, and several specialist types are available today, only three major ones are now in common use: the dynamic or moving coil microphone, the condenser microphone, and the piezoelectric microphone.

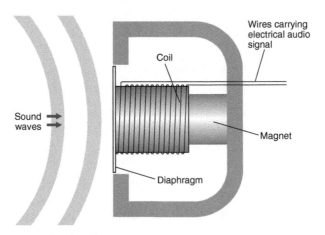

18. Dynamic microphone.

In a *dynamic microphone* (Figure 18) the diaphragm is attached to a coil of wire that surrounds a stationary magnet. A small voltage is induced in the coil when it moves, which in turn gives rise to a weak current. The low voltage means that the quality of the response is too poor for measurement work, so dynamic microphones are mainly to be found in concerts and recording studios.

In a *condenser microphone* (Figure 19), the diaphragm forms one plate of a capacitor (formerly called a condenser, hence the name). A capacitor is a pair of parallel metal plates with a thin layer of either air or some other material which does not conduct electricity (known as a dielectric) between them. Attaching one plate to the negative pole of a battery charges it with electrons. All metals contain free electrons, and those in the other plate are repelled by the electrical field arising from the large number now residing in the negative one. The repelled electrons flow from that plate, to leave it positively charged. The whole capacitor, therefore, now has a voltage (called a polarizing voltage) across it, and the microphone is ready for use: the diaphragm plate moves in and

70

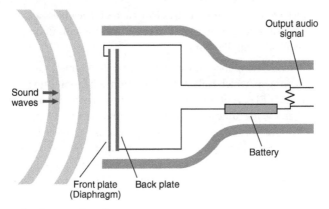

19. Condenser microphone.

out with the compressions and rarefactions of the sound waves
that impinge on it. A large electrical resistance stops the charge
from rapidly escaping, so instead the sound waves are transformed
to patterns of voltage ripples. Condenser microphones have an
excellent frequency response and are used as measurement
microphones in laboratories and in SLMs. They also respond
more rapidly to sudden sounds than do dynamic microphones.

Crystal microphones and *ceramic microphones* exploit the
piezoelectric effect, in which quartz or some other crystalline
material produces a voltage when slightly compressed. Most
landline telephones use these, as do call-centre headsets.

Other types of microphone, of historical or specialist interest,
include the following:

Electret microphones are made of permanently charged
material (the electrical equivalent of a magnet). They function
in a similar way to condenser microphones.

MEMS (microelectromechanical systems) microphones are
condenser microphones etched directly on to silicon chips.

71

Just a few square millimetres in area, their cheapness and robustness means they are used in mobile phones, among many other applications.

Optical microphones use a shiny silicon membrane as a diaphragm, which reflects the light from a light-emitting diode (LED). A photo detector measures changes in the light when the membrane is vibrated by sound waves, and an electronic circuit converts these changes into an electrical signal. Such microphones are compact and robust and are unaffected by local electromagnetic fields, so they are used, for example, to allow communication between patient and staff during MRI scans.

Pressure zone microphones (PZMs) are designed for use near a hard reflective surface (such as when placement directly on a stage floor is required). With a conventional microphone, reflection from the surface would interfere with the direct sounds to the microphone, but a PZM overcomes this by having its diaphragm so close to the surface that most wavelengths overlap.

Ribbon microphones use a strip ('ribbon') of metal rather than a diaphragm, and unlike the eardrum and most other kinds of microphone, they respond to the pressure difference between the ribbon's sides, which mean that velocities with which air molecules in the sound wave move are detected, rather than sound pressure patterns. Such microphones are often used by commentators in noisy surroundings, because a sound which comes from all directions will impact both sides of the ribbon simultaneously, so that it will not respond. For this application, the microphones have a projection that is held to the upper lip, which helps guide the commentator's voice so that it impinges on one side of the ribbon only.

An alternative for capturing speech in noisy environments is the *lavalier microphone*, which is a small clip-on electret or dynamic microphone. Lavaliers have the advantage that they can readily be concealed under clothing.

A *sound intensity probe* consists of a pair of face-to-face microphones which measure the pressures of different parts of the same sound wave. From these the molecular velocity and hence the sound intensity is calculated.

An important criterion in choosing a microphone is directionality: an ideal omnidirectional microphone is equally sensitive to sound from any direction, and is required to capture a full soundscape, for example. A unidirectional microphone, which picks up sound from one direction only, is ideal for picking up speech or song in noisy environments.

Except for the ribbon microphone and sound intensity probe, every type is naturally fairly omnidirectional in response, providing that the wavelengths of the sound waves that impinge on it are larger than the diaphragm. In order to make a microphone sensitive mainly to sound from immediately in front or behind (bidirectional), all that is needed is to allow the back as well as the front to be open to the air. The diaphragm will not move very much in response to sound waves arriving from all around it, because the rise and fall of pressure at the front would be very similar to that at the back. But any waves incident on the front or back face of the microphone will be picked up readily.

A simple way to achieve directionality is to mount a microphone at the focus of a reflector whose cross section is a parabola. This shape reflects incident sound waves on to the diaphragm, as long as they are smaller than the reflector. Alternatively, a microphone can be mounted at the end of a tube with slits along its sides, to make a *shotgun (boom) microphone*. Sound waves passing straight down the tube travel unimpeded to the microphone, but those from other directions enter through the slits. Each such sound will enter multiple slits, so many versions of it will be formed, each with a different phase. The versions will therefore largely cancel each other out through destructive interference. Shotgun microphones are widely used for outdoor recording, often accompanying cameras. They are, however, highly frequency dependent.

Electricity to sound: loudspeakers

Loudspeakers are microphones in reverse: if a dynamic, condenser, crystal, or ceramic microphone is supplied with a varying current, it will vibrate to produce sound waves (such microphones are therefore referred to as reciprocal transducers). Most actual loudspeakers are dynamic microphones in reverse, and are called moving coil loudspeakers (Figure 20). As the name implies, an electrical signal is fed to a coil (the voice coil) attached to a diaphragm which is usually in the form of a cone. The coil surrounds a magnet and the electromagnetic field set up in the coil by the signal causes it, and the diaphragm, to move.

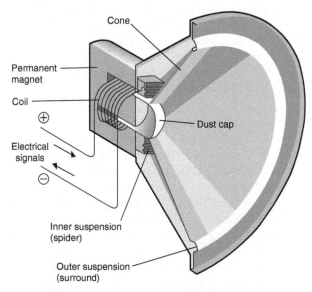

20. Loudspeaker.

Loudspeakers are intrinsically inefficient devices: most of the electrical energy that passes into them appears as heat, with only about 1 per cent being audible. Hence, amplification is vital. Transistors make amplification a simple matter today, the main challenge being to ensure that each frequency is amplified to an appropriate extent, which, given the non-linear attributes of the hearing system, means that different degrees of amplification must be applied to different frequencies if the pitch envelope of the output sound is to remain unchanged. It is essential to deploy them with care. Quite apart from the deleterious effects on hearing, loudspeakers can be easily damaged, especially by artificially produced sounds, which may have an extremely fast rise time. Also, if the microphone picks up the sound of the loudspeaker to which it is sending its output, a positive feedback loop is set up, resulting in an all too familiar squeal.

Loudspeakers are the least hi-fi link in the music production chain: though simple in principle, their design presents many practical challenges. The voice coil must return to exactly its start position when the signal falls to zero, with no oscillation beyond it, and yet must be free to move. The cone must hold its shape while it vibrates, be very light, yet rigid enough not to sag under gravity; large enough to move large volumes of air at low frequencies (where only powerful waves are audible), yet small enough so that they can move back and forth over 10,000 times a second at high frequencies. Also, the casing must not resonate at any frequency.

In practice, it is much easier to use speakers in groups, usually all in the same unit—a small tweeter for frequencies above 2 kHz, a larger mid-range speaker (50 Hz to 5 kHz), and a woofer (30 Hz to 800 Hz). For those who love lots of bass, there may also be a subwoofer (20 Hz to 200 Hz).

Subwoofers are usually *active* loudspeakers, which means they contain their own amplifiers (and hence need their own power supply). Most other speakers are passive, driven by a signal which has already been boosted by an amplifier within the hi-fi (or other) system.

A loudspeaker without a case is almost silent, for the simple reason that the high-pressure pulses generated at its front simply slip round to the back to fill the low-pressure area that has just formed there. Hence, loudspeakers may be mounted in an airtight box. If the box is small, however, the air in it resists being compressed as the diaphragm moves in. An alternative solution is to place the diaphragm in the centre of an annulus called a baffle. The baffle must be large enough such that, by the time the pressure pulse has travelled round it to the back of the diaphragm, the low-pressure area there has gone (or, to put it another way, the distance the sound waves must travel is longer than a quarter of a wavelength for the longest waves—the lowest frequencies—of interest).

Helmholtz resonance can be used to extend the low-frequency performance of a loudspeaker. A hole (*port*) is made in the front of the box in which the speaker is mounted, and the cavity then resonates at low frequencies. If the resonant frequency of the box lies below that of the speaker, a pressure pulse produced when the speaker diaphragm moves back (the backwave) will make its way within the box to emerge at the port, where it will be in phase with a new pulse just produced by the front of the diaphragm. These two in-phase pulses will reinforce each other. This applies to pulses which make up waves of any frequency that falls between the resonance frequency of the diaphragm and that of the box.

One disadvantage is that the puffs of air from the port can sometimes be heard, and another is that sounds are less crisp because each signal is followed by a short fading 'tail' of resonances. Also, pulses forming sound waves with frequencies below that of the port are cancelled out by the subsequent wave from the diaphragm. Frequencies above both resonances are neither enhanced nor reduced.

Our brains are so good at filling in gaps in sounds that we can exploit this to make do with quite basic loudspeakers. In nature, a set of tones of 200 Hz, 300 Hz, 400 Hz, and 500 Hz will almost always be harmonics (overtones) of a fundamental at 100 Hz. Because the brain's hearing centre (see Figure 16) 'knows' this, it will confidently decide that the 100 Hz is actually there. But, if the set of tones is coming from a small loudspeaker, it's likely that there will not in fact be a 100 Hz tone present. This effect is known as the missing fundamental, and is why the lack of low frequencies in the Epidaurus theatre (Chapter 1) did not sound strange: the hearing centres of the audience interpolated them. It also explains the relative clarity of the earliest telephones, which did not transmit low frequencies well.

The introduction of effective microphones, amplifiers, and loudspeakers revolutionized the record industry—and this revolution could be rapid since the records themselves did not need to be changed. Social changes soon followed: quite suddenly, almost anyone could access music, and select what they listened to as well.

According to sound studies expert Jonathan Sterne, such new sound media 'called into question the very basis of experience and existence'. And the effects on performers were profound: as music historian Robert Philip points out, many were aghast at the number of errors they heard in their recorded performances. Musicologist Mark Katz suggests that such performers became trapped in a 'feedback loop' in which they attempted to produce more and more 'perfect' performances, only to be disappointed all over again when they listened to recordings of the result. As a result, performance became less individualized, more standardized, and lost spontaneity. The experience of listening to recordings caused another kind of feedback too: violin vibrato, for example, was originally a phonographic effect, but was soon being imitated by performers.

The next revolution in this field was the invention of stereophonic phonograph recording in 1933, achieved by recording the two channels on the two walls of the groove, at 90° to each other and 45° to the vertical. The introduction of stereo records meant that in principle the whole three-dimensional sound field of the original performance could be recreated, giving rise to questions of ideal speaker placement which have fascinated music buffs to this day. It also gave rise to the concept of 'fidelity', since what one was now doing was recreating an original performance. For a while at least—with advancing post-processing and mixing techniques, by the 1970s many pop pieces were never performed as such; much of their content was added after the band had gone home. For classical music however, faithful recording has remained key. Nevertheless, despite decades of interest by many millions of amateurs and professional listeners, recorders, performers, and players, 'fidelity' is still unquantifiable.

Storing sounds

Once the technology of amplification and loudspeaker design had been perfected, the main concern was the fragility of records. A whole culture developed concerning how to handle, house, clean, and—once correctly trained—play them. The auto-changer came as a great relief to some, though to others was regarded as either a spoilsport or an incipient record-damager. Partly because of this mystique and partly because of the high quality of some of their covers, vinyl records, and especially LPs (later called albums), were venerated as objects in a way that no other recording media have ever been, and a few are still sold to this day.

In 1964 analogue cassette tapes were introduced as a robust and compact alternative to discs, and to begin with were very popular: although recorded ones were readily available, many preferred to buy a vinyl record and (illegally!) transfer it to cassette to listen to while keeping the record pristine. Radio broadcasts could also be recorded in this way and the music centre, comprising radio, cassette recorder, and record player, became popular since it allowed cassettes to be recorded with a minimum of fuss.

There are two major disadvantages with tape, however: track selection requires time-consuming winding, and high-frequency hiss is unavoidable. The latter was reduced somewhat by the many variant Dolby systems on offer, all of which work by recording a version of the track with the high frequencies boosted, and then suppressing them on playback. This is an example of a technique called companding (a portmanteau of compressing and expanding).

Just as stereo spawned new interests and behaviours, so did tape. Mixer tapes were one outcome, and another was the Sony Walkman, which at last allowed music lovers to listen to their music wherever they happened to be, while only annoying people sitting next to them.

But what was needed was to do away entirely with analogue recordings, that is, those in which a sound is stored as a continuous varying pattern (whether physical, as in a vinyl record, or magnetic, as in a cassette tape). In a digital system, the recorded signal is encoded as a string of numbers, and once in that form it can be stored, transmitted, or copied with neither degradation of the signal nor increase in background noise.

At first glance, it may seem that in order to capture the intricacies of a complicated sound wave (like those in Figure 9), the amplitudes of a great many points of that wave must be measured and coded. In fact, the sound need only be sampled twice as frequently as the highest-frequency component that one wishes to preserve. So, to encode all frequencies in a signal up to 8 kHz, one must sample at 16 kHz (this is known as the Nyquist theorem). If one samples at a lower rate, the encoded data become distorted, an effect known as aliasing.

With the introduction of the compact disc (CD) in 1982, a wholesale shift from analogue to digital began. On a CD, digitally encoded signals are stored as patterns of dark pits in a shiny metal layer, which are scanned by a laser. The laser reflects from smooth areas but not from the pits; the CD player interprets the reflections as 1s and the non-reflections as 0s, and the strings of 1s and 0s encode the audio information as a sequence of binary numbers. Originally, CDs were claimed to have the impressive though not especially useful capability to play even if coated with marmalade. The claimants neglected to mention that the marmalade had to be applied to the upper side of the CD, as the coding is on the underside.

Nowadays of course, music is routinely bought, stored, and played without using physical media—audio files can simply be downloaded to a computer and played through a wide range of output devices. Often, the computer is part of an MP3 player (MP3 is short for 'Moving Picture Experts Group-1 or Moving Picture Experts Group-2 Audio Layer III').

The magic of MP3 audio files is how small they are: about one-tenth the size of equivalent CD files, which means a minute of MP3 music can be squeezed into a megabyte. This impressive reduction is achieved partly by a technique called Huffman coding, in which the symbols that appear most often are coded in the shortest possible way, and partly by coding more fully those frequency bands which people will notice most if they are disrupted (mainly speech frequencies), and providing only sketchy versions of those frequencies of less concern to us.

Because MP3 players take into account both the music and the listener in deciding what information to leave out when the song is compressed, Sterne has concluded that:

> the MP3 carries within it practical and philosophical understandings
> of what it means to communicate, what it means to listen or speak,

how the mind's ear works, and what it means to make music. Encoded in every MP3 are whole worlds of possible and impossible sound and whole histories of sonic practices . . . . MP3 encoders build their files by calculating a moment-to-moment relationship between the changing contents of a recording and the gaps and absences of an imagined listener at the other end. The MP3 encoder works so well because it guesses that its imagined auditor is an imperfect listener, in less-than-ideal conditions. It often guesses right.

Efficient coding of music exploits the fact that, over billions of years, our hearing systems have evolved to respond to the sounds which are of most relevance to us. This, combined with limitations due to the nature of sound, restricts our immediate access to the world of sound to a frequency range which is only a thin slice of what is actually out there. Those unhearable realms are the subject of Chapter 6.

Chapter 6
Ultrasound and infrasound

Bat sounds

Many textbooks will tell you that you can hear from 20 Hz to
20 kHz. Don't believe them: if you are over twenty, you are probably
deaf to sounds above 17 kHz. The high-frequency limit of our
hearing declines so significantly and so predictably with age that a
youth-repelling generator of higher-frequency sound called the
Mosquito has been used by irate shopkeepers since 2009, as only
teenage ears (and those of children and babies) are still capable of
detecting the annoying tones the device produces. However, it is
probably fair to say that we who cannot hear anything about
17 kHz are not inconvenienced by the fact, showing that it is of
little evolutionary benefit to hear such high notes, otherwise we
would have developed more robust hearing systems.

Nevertheless, we *are* missing out—not because the ultrasonic
soundscape is an especially rich one, but because we cannot
exploit some handy physical properties of sounds that are
negligible at those frequencies that we can hear. The bat, however,
is not so handicapped.

The sophistication with which bats employ ultrasound is
astounding. In pitch darkness, an Egyptian fruit bat (*Rousettus
aegyptiacus*) with an 80 cm wingspan can easily fly between vertical

rods just 53 cm apart without touching them. To achieve such feats, and to hunt, bats use echolocation—generating ultrasounds and timing the delays until they hear the echoes—which inform them of the distances of nearby objects.

Sightless humans approximate this, judging distances to walls from echo delays, but what they end up with is not a vision of the world. It cannot be, at the wavelengths they can hear: imagine a jetty standing in a calm lake. A light breeze blows, forming ripples about a centimetre in wavelength, which bounce off the jetty's columns in circular patterns (Figure 21). Later, a gale sets in, making much longer waves, which roll past the columns as if they're not there. For waves of sound, the same principle applies: they are only affected by (and hence can only detect) obstacles larger than they are.

So, if a bat tried to echo-locate a moth using an audible tone—middle C, let's say, which has a wavelength of 132 cm—it would only succeed if the moth were well over a metre across. To detect a 1 cm moth, sound waves of at least 33 kHz are required. In fact, bats typically produce sounds of 80 kHz (smaller species produce higher frequencies, and the full range is about 40 kHz to 120 kHz). Such high frequencies allow them to deal not only with isolated objects, but with complicated three-dimensional distributions of twigs, leaves, and insects, many in relative motion to each other—and all in relative motion to the bat.

But it would do the bat no good to generate a continuous ultrasonic whine: the result would be a bewildering montage of overlapping, interfering echoes, impossible to decode. What bats need are pulses of sound, and they must be brief: if two moths are a centimetre apart, then, to produce separate echoes, a sound pulse must have passed beyond the closer one before reaching the second, and hence must be less than a centimetre in length, and therefore less than 30 microseconds in duration.

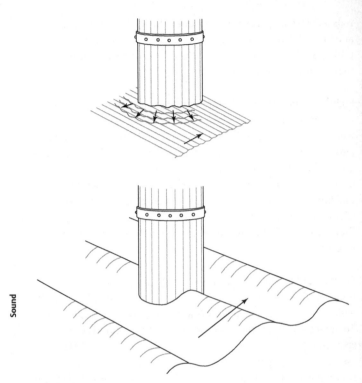

21. Ripples vs waves.

It is impossible for bat or machine to produce a short-duration pulse with a single wavelength, unless that wavelength is much shorter than the pulse itself. Instead, bats produce very sharp clicks. Thanks to Fourier analysis, we know that such a click is equivalent to a mixture of waves of different frequencies (Figure 12), and the sharper the click (that is, the quicker it rises from silence to maximum), the wider the range of frequencies produced—and thus the more accurately distances can be judged.

So is the ideal sound for a bat an extremely short-duration click? Not really: a short click means a low-energy one, which cannot travel far without fading. Humans faced this problem themselves in designing radar and sonar systems, and cracked it in 1960 with the invention of frequency sweep. This utilizes a relatively long pulse of gradually increasing frequency. The fairly long duration allows plenty of energy and hence range, and the changing frequency means that echoes from objects at different distances are distinguishable by the differences in their frequencies.

When bats use this frequency sweep technique, the pulses last about 2 to 3 milliseconds, during which the frequency falls (rather than rising as in the human version) by about an octave. The more such pulses a bat makes, the more information it receives, and so it modifies its click rate depending on the challenges it faces, from around ten pulses a second when cruising to as high as one hundred in a half-second period, when the surroundings become complicated or when nearing prey.

As discussed in Chapter 2, if the object from which a sound echoes is in motion (relative to the sound source), the frequency of that sound will be shifted by the Doppler effect. Both human and bat sonar systems exploit this to determine the velocities of such objects, with the distinction that, while our systems measure the change in frequency, bats modify the frequencies of their output pulses until the echoes they hear are the same as they would be, were the target stationary (so, for example, if a bat is approaching an object, the frequency of the echoes from that object rise, so the bat reduces the frequency of its output by an amount that brings the frequency of the echoes back down to match the bat's original frequency).

A different kind of sound

Another physical difference between ultrasound and the audible range is that ultrasound readily forms beams, which is of considerable advantage to bats. An 80 kHz tone passing through a 1 cm opening will form a conical beam which spreads to about

90 cm wide at a 1-metre distance (Box 9). In bat species which project their ultrasound through their nostrils, interference effects between the two sources mean that the beams are narrowed further. Not only does this concentrate the acoustic energy, so allowing greater distances to be probed, it also reduces the number of distracting side-echoes. A 2015 study of bats approaching drinking pools showed that the bats opened their mouths wider when they were close to the water, presumably for this latter reason.

Evolution could no doubt have furnished bats with generators and detectors that work at even higher frequencies, but the absorption of such sounds by the air is an insuperable barrier. Experimentally it has been determined that, at the sort of conditions bats love best (25°C, 50 per cent relative humidity), 100 kHz sounds are absorbed by 3 dB per metre: that is, they fall to about half of their original intensity over that distance. Conversely, 30 kHz sounds are absorbed only at the modest rate of 0.7 dB per metre, which means their intensity falls by about 15 per cent. (The increase of sound absorption with frequency is the main reason why thunder sounds less crackly and more booming as distance increases.)

Why does this happen? Air is composed of molecules, all randomly drifting around at a range of velocities and frequently bouncing off each other. On a hot day, these velocities increase—in fact, temperature is just a measure of the velocity of a large group of molecules (Box 10). A sound wave is a sequence of alternating high and low pressures moving through the air, so at any given

> **Box 10**
>
> Absolute gas temperature (that is, as measured in kelvins) is proportional to the kinetic energy $\frac{1}{2}mv^2$, molecular mass m, molecular velocity v.

location reached by the sound, the air molecules will first bunch up close together and then spread apart, before bunching together again. When molecules bunch, they slow down, just as a people hurrying through a crowded station concourse in different directions go slower the more of them there are. When this happens, the energy of the molecules changes form: although they move around more slowly (their kinetic energy falls, in other words), they spin and stretch more (so, their internal energy increases).

An analogy might be to pairs of cricket balls linked by strong steel springs—a fairly accurate model of a diatomic (2-atom) molecule, like the molecules of nitrogen (N_2) and oxygen (O_2) which together make up 99 per cent of our atmosphere. As soon as the compressional part of the sound wave has passed and the molecules move apart again, they spin and stretch less, and move around faster once more. Play middle C, and the sound waves produced force the energy to switch back and forth between the kinetic and internal forms 262 times a second.

But, if the frequency is sufficiently high, the time available becomes so short that a molecule cannot complete the conversion of internal energy to kinetic energy quickly enough before it is time to reverse the process. As a result, the velocity of sound falls and the sound wave dies away rapidly. The actual frequency of the sound wave at which these changes begin depends on the medium, being far higher in solids and liquids than in gases. Other properties of the medium, especially viscosity, also contribute to this effect.

This phenomenon is very useful to us: as the ultrasound wave fades, its energy spreads through the medium, heating it up. This heating effect has many applications, including the internal warming of body tissues to improve blood flow or to treat damaged muscles and joints.

Medical ultrasound

Ultrasound has many other medical uses too, both in the areas of scanning (almost everyone in the developed world is now scanned before birth) and treatment (like the removal of dental tartar, where 25 kHz sound is used in conjunction with a water jet). Unlike many other medical treatments, ultrasound can be switched on and off instantly, often requires low-cost technology, and usually involves minimal patient preparation. The fact that ultrasound generators are relatively portable and need little ancillary equipment means that they can be used outside the hospital environment, from diathermy (deep-tissue heating) units found in many gyms to wound cauterization systems, deployed on battlefields to save the lives of soldiers who would otherwise perish from blood loss.

Ultrasound has been used for the treatment of tumours of many kinds—including otherwise inoperable brain cancers—by a technique called either HIFU (high-intensity focussed ultrasound), or HITU (high-intensity therapeutic ultrasound). As well as destroying cancerous tissues by heating them up (to about 90°C over regions about the size of a grain of rice), ultrasonically induced bubble formation in tumours has been used to make them more susceptible to chemotherapy. The main challenge in the use of ultrasound for treatments like these, which need precise and accurate targeting, is that the paths of the ultrasound beams depend on the densities and elasticities of the intervening body tissues. So, models of body parts made from artificial tissue-mimicking materials are used to calibrate and programme the equipment.

A more straightforward use of ultrasound in medicine is lithotripsy, in which high-power pulses simply pound kidney stones *in situ*, smashing them to pieces small enough to pass out with urine.

Scanning with ultrasound

One of the best-known uses of ultrasound is the scanning of foetuses. Since the diaphragm of a conventional loudspeaker could not move quickly enough to produce the megahertz frequencies required, a piezoelectric projector is used instead. Gel is applied to the abdomen, so that there is no air layer to reflect or absorb the sound. The ultrasound waves bounce off interfaces between media with different impedances, such as bone/muscle, or skin/amniotic fluid. Hence, by measuring precisely how long it takes for echoes to return from these interfaces (taking account of the differing sound velocities in each medium), their positions can be worked out. By moving the beam around, a detailed three-dimensional map can be calculated, and converted to a real-time video image.

Like most loudspeakers, a piezoelectric transducer is reciprocal—supplied with an oscillating current it produces a sound wave, and, conversely, when struck by a sound wave it generates an electrical signal. So, in a foetal scanner the scanner head acts as both the source of the ultrasounds and the detector of their echoes.

With very high-frequency sound beams relatively sharp images can be made: a 1 MHz (one million hertz) signal can image details at the millimetre scale, the exact value depending on the acoustic properties of the tissue bring viewed, and many scanners now go up to 15 MHz, or 50 MHz for eye and skin scanners.

But even this is paltry compared to the eight *billion* hertz signals generated by an acoustic microscope, which can consequently resolve details as small as 0.03 micron. Unfortunately, such high-frequency sounds are absorbed by almost all media before they have travelled a single millimetre, and the only useable exception is liquid helium—which will boil if not kept colder than 5 kelvin (−268°C). The elaborate cooling systems required mean that acoustic microscopes are not cheap: the reasons they are used at all are that they can probe under the surfaces of samples, and that some materials which are hard to distinguish from their surroundings visually are highly reflective to sound.

Outside the medical field, one of the commonest diagnostic applications of ultrasound is in the detection of flaws and cracks—in railway lines, for instance. To locate these, a series of sound bursts is sent along the object to be tested. In pulse-echo mode, the transmitter and receiver are positioned together: if a flaw is present, the pulses are reflected and their arrival times indicate the flaw position. In transmission mode, the detector and transmitter are separated and any changes in the pulses in transit indicate the presence of inhomogeneities in the test object. This approach can also be used for the determination of mechanical stresses in solids: because the elastic moduli of materials alter when stressed, their sound velocities also change locally.

The power of ultrasonics

Though thermal effects of high-power ultrasound have many applications, it can also deliver its energy through mechanical effects on the medium. The snapping shrimp provides a rare natural example of the use of the shocks and stresses made by ultrasound (together with audio frequencies) to kill. The loud and sudden click produced when the creature snaps its claw includes frequencies up to 200 kHz, powerful enough to kill or stun both prey and would-be predators alike.

Whether utilized by shrimps or humans, the mechanical power of ultrasound is usually delivered through cavitation, which is the formation and violent collapse of tiny bubbles. Almost all liquids contain such bubbles, made either of their own vapour or of air. When the pressure of a liquid falls, these bubbles grow (which is the cause of the frothing when a pressurized container of fizzy drink is opened). Since a sound wave is a sequence of low and high pressures, it causes bubbles to swell and shrink rapidly, and at high powers and frequencies the bubbles pulsate so violently they disrupt and implode, suddenly releasing their vibrational energy in the form of heat. The temperatures involved may exceed those on the surface of the sun, and can make the liquid glow with light (a phenomenon called sonoluminescence, which snapping shrimp also initiate).

Because the energy appears only within tiny volumes, the body of liquid does not get particularly hot. But the bursts of highly concentrated energy can be used to initiate chemical changes (sonochemistry), and to clean and sterilize submerged objects such as medical instruments. Cavitation-based cleaning is most effective in the 20 to 50 kHz range. At higher frequencies, ultrasound also causes agitation of liquids, which dominates cleaning effects in the range 100 kHz to 1 MHz. In practice, ultrasonic cleaning baths use both effects.

High-power ultrasound (without cavitation) is routinely used for fluxless soldering of printed circuit boards, in which the electrically heated tip of a soldering iron is vibrated at ultrasonic frequencies. At still higher powers, fine wires can be welded together, heated from within by the friction caused by ultrasonic vibrations. A great advantage of this is that, since the heating effect is restricted to the insonified (sound-filled) wires, the ultrasound does not heat up nearby components. Many other materials are welded ultrasonically; the frequency chosen depends on the size of the parts to be welded, from 60 kHz for the smallest elements to 10 kHz for the largest.

High-power ultrasound can even be used to levitate small objects. Although the forces involved are feeble, in microgravity environments such as the interiors of space stations they could be used to hold delicate instruments in position during assembly, or to prevent highly reactive chemicals from coming into contact with anything.

The outer limits: phonons

At extremely high frequencies, sounds behave in ways radically different to those with which we are familiar: just like electromagnetic radiation, in which the highest frequencies behave much more like particles than waves (hence the individual clicks of a Geiger counter in registering the presence of gamma rays), the highest-frequency sounds behave like particles called phonons. Their existence was discovered indirectly: by the late 19th century, it was known that a specific amount of heat was needed to raise the temperature of a substance by one degree. The specific heat of water is higher than that of, say, oil, which means it takes longer to boil a kettle of water than a kettle of oil.

Gases and solids have specific heats too, but there was a puzzling anomaly: increasing the temperature of a solid takes about twice as much heat energy as raising the temperature of the same amount of the same substance in its gaseous form. This means that solids (and certain liquids) must have a means of storing heat which is unavailable to gases. The mechanism is vibration: a molecule in a solid can oscillate around its equilibrium position, like a pendulum. But, unlike a pendulum, an oscillating molecule cannot slow gradually. The laws of quantum mechanics mean that it must jump from a rapid to a slow oscillation, and when it does so it transfers a phonon of vibrational energy to another molecule. The ways in which solids conduct heat and electricity can be accounted for by the behaviour of phonons.

Infrasound

As the lower frequency limit of human hearing is approached, the intensity of just-audible sounds increases: a pure tone that is just audible at 20 Hz is nearly 300 million times more powerful (all other things being equal) than a just-audible tone at 4 kHz. Such powerful low-frequency sounds are rarely encountered in air, but solid-borne versions are commonplace on construction sites, in underground stations, near motorways, and in seismically active areas, where they can easily be felt.

Low-powered airborne infrasounds, on the other hand, surround us all the time. They are even generated when we walk: the cyclic air pressure changes at our ears due to our up and down head movements constitute an infrasound wave at around 1 Hz. Sea waves generate infrasounds at around 0.2 Hz. The lowest-frequency natural sounds of all come from high in the air and deep underground: both aurorae and volcanoes produce infrasounds at around 0.01 Hz.

One of the key characteristics of infrasound is that it can travel far further than audible sound, through sea, ground, or air; in air, infrasounds from thousands of kilometres away (made, for example, by erupting volcanoes) can easily be detected, though not usually by microphones. Just as we feel, rather than hear, infrasound, so we use specialized barometers to measure it. Infrasound can also be detected by means of the temperature changes it causes.

There is evidence that infrasound has a range of effects on humans not shared by other types of sound, including enhancement of emotional responses: when a concert of modern classical music was accompanied by infrasound, there were far more people who hated the music, and far more who loved it, than at a concert of the same music with no infrasonic accompaniment. Other experiments have shown that drivers exposed to infrasound can get very tired very quickly, and it has even been suggested as a cause of supposed hauntings, partly through direct emotional effects and partly through vibration of the eyeball, giving rise to visual disturbances.

Infrasound has proved to be a reliable means of detecting bolides, which are meteoroids that explode in flight. Since they travel supersonically through the atmosphere, they generate infrasound-rich sonic booms as they fall. More infrasound is generated when the bolides explode. All these airborne infrasounds generate solid-borne versions when they reach the Earth, and surface waves are made when the bolide fragments crash to Earth. The frequencies, timing, and amplitudes of all these sounds can be combined to give a detailed analysis of the path, motion, and energy of the bolide.

However, infrasound is usually of more consequence to us when it travels through the Earth—and ultrasound really comes into its own underwater. It is to these media that we turn in Chapter 7.

Chapter 7
Sound underwater and underground

Second sight

If our world had always been blanketed in fog, we would probably make little use of sight, and rely on our ears instead. If the fog proved opaque to radio waves, our telecommunication systems would no doubt have taken an acoustic turn, too. The ocean is a world of just this kind: visibility is poor due to the many suspended particles in the water, radio waves quickly die. And the water is full of sound: fish, marine mammals, human divers, submarines, and underwater robots all use it for subtle and complex forms of sensation, communication, and sometimes attack.

All this would be news to a scientist living only a century ago. It's true that Leonardo da Vinci pointed out around 1490 that 'If you cause your ship to stop, and place the head of a long tube in the water, and place the other extremity to your ear, you will hear ships at a great distance from you', but no one took much notice of him.

It was not until World War I that listening to the underwater sounds of vessels was attempted, and not until World War II that the richness of the underwater soundscape became clear. This happened partly through two bizarre events. In 1942, acoustic buoys in Chesapeake Bay, which had been deployed to detect German submarines, all alerted at once. A flotilla of destroyers enthusiastically depth-charged the area—but no tell-tale oil slicks appeared and the only victims were thousands of fish. Later that same year, all along the western coasts of the US, every acoustic mine (set to protect ports from anything with a propeller, in fear of a Japanese invasion) detonated at once—and again, the sole result was a multitude of dead marine life.

If the US military had known a little more about a fish called the croaker, much embarrassment and many dead fish could have been avoided. For all its undistinguished appearance, this small brownish creature has a very loud voice indeed—something like a magnified woodpecker—and croaker colonies give voice every dawn, all together, much like birds.

Noisy fish were an almost complete surprise to biologists. There was a hitherto unquestioned belief that the depths were as silent as they were proverbially said to be. In fact, while the oceans had been routinely exploited for coastal and international travel and as a source of food, what lay beneath the surface was largely unknown and apparently rarely even speculated upon, beyond wild tales of sea monsters and legends of sunken cities. The dark underwater world seems to have been regarded as wholly alien and no place for people—even sailors regarded the depths as forbidden territory and most of them did not even learn to swim. While artists and writers—and psychologists too—often viewed the sea as the wild domain of raw nature, their attention was confined to its rough surface alone. Beneath that surface was a negative region, assumed to be as devoid of sound as it was of light.

However, from the early 1940s, listening with hydrophones (underwater microphones) revealed just how noisy the 'silent deep' really is, with a cacophony of sounds extending well above and below the human frequency range, from a vast multitude of sources, most then unidentifiable. The loudest source in some areas turned out to be colonies of snapping shrimp. In fact, the upper layer of the sea is never quiet, with a background generated by breaking waves, rain, and lightning as well as by marine life. In shallow areas, swirling sediment is an additional, omnipresent source of sound.

Hearing underwater

There are two reasons for the apparent silence of the sea: one physical, the other biological. The physical one is the impedance mismatch between air and water, in consequence of which the surface acts as an acoustic mirror, reflecting back almost all sound from below, so that land-dwellers hear no more than the breaking of the waves.

Submerge your head, and the biological reason manifests itself: underwater, the eardrum has water on one side and air on the other, and so impedance mismatching once more prevents most sound from entering. If we had no eardrums (nor air-filled middle ears) we would probably hear very well underwater.

Underwater animals don't need such complicated ears as ours: since the water around them is a similar density to their flesh, sound enters and passes through their whole bodies easily, without the need of pinnae to coax it in, nor of drums or windows to transfer it from one medium to another. Fish do have ear bones, called otoliths. They are made of calcium carbonate, the high density of which provides a sufficient impedance difference to allow vibration by sound waves. This motion is transferred to stereocilia growing on hair cells, which, like ours, send nerve signals to the brain. Other patches of hair cells, called neuromats, are distributed on the skins of fish.

Two other structures augment the hearing of some fish. The first, the swim bladder, is an air-filled sac which functions like the ballast tanks of a submarine, changing the buoyancy of its owner as required, so that the fish can sink or rise without muscular effort. The swim bladder vibrates readily to sound waves, and functions as a fairly sensitive hearing organ up to about 3 kHz. It has, however, a great disadvantage: being a single, symmetrical organ, the swim bladder provides no information about the direction from which sounds arrive.

It is supplemented by a second structure: the lateral line, a fluid-filled tube that runs along the sides of the fish and functions as a direction-sensitive sound detector at low frequencies (~160 Hz to 200 Hz). Unlike the stereocilia in our ears, which wobble when sound waves induce vibrations in the basilar membrane to which they are attached, lateral line stercocilia are pushed and pulled directly by incoming sound waves, which means that the direction of the sounds are sensed directly (and also means that fish sense the motion of water molecules, rather than the pressure changes that we do). This makes fish very hard to creep up on.

Techniques and transducers

The earliest serious technological use of sound underwater was the sounding bell system, in which underwater bells placed near ports could be detected by ships fitted with primitive hydrophones in the form of carbon microphones in waterproof casings. A watchman listened on the ship in stereo and could guide the ship to port even in poor visibility. This system was fitted to many ships in the period 1875–1930, including the *Titanic* and *Lusitania*. By 1923, there were thirty bells around the UK coast. From the 1910s, the system was gradually replaced by echo sounding, which involves making a sound underwater, timing its echo, and working out the distance from a knowledge of the velocity of underwater sound: another example of the pulse-echo technique.

From echo sounding, modern sonar (sound navigation and ranging) systems developed. In an active sonar system, short-duration sound impulses ('pings') are projected from a vessel to echo from objects in their path. As well as determining distance, changes in the frequency of the received pulses are used to calculate the relative velocity of the source (using the Doppler effect).

Passive sonar systems simply listen for underwater sounds, especially those of shipping. Automatic recognition techniques allow different types of shipping to be identified from their engine sounds or even from the humming of their electrical systems: each vessel, in fact, has its unique signature or *acoustic fingerprint*. This technique became very important in the Cold War for recognizing and tracking enemy ships and submarines. Active and passive sonar systems are sometimes deployed on buoys (*sonobuoys*), which are equipped with radio systems to report their findings.

The hydrophone is the key instrument for underwater acoustic work; almost all those in use today are piezoelectric, usually based on a synthetic ceramic called PZT (lead and titanium zirconate). Unlike microphones, hydrophones must sometimes be very large, in order to achieve directionality at low frequencies. For this reason, the sides of some submarines are almost completely plated with hydrophones. These flank arrays, as they are called, are usually made of polyvinylidene difluoride (PVDF).

The underwater equivalent of the loudspeaker is known as a projector. Projectors have a limitation that does not apply to loudspeakers: when the surface moves inwards to create the rarefaction phase of a sound wave, cavitation results if the rarefactional pressure is lower than that of the surrounding water. The bubbles scatter and absorb the sound, silencing the projector. The greater the depth, the higher the water pressure becomes, so the more sound power the projector can produce before the onset of cavitation.

For high-power applications, such as geophysical surveying for oil and gas exploration, airguns are used to produce underwater sound. In these, a small cavity is charged with compressed air and a relay releases the pressure suddenly, rapidly forming a large air bubble and an accompanying burst of sound. The pulse frequency is around 20 Hz to 200 Hz and the amplitude is very high (probably the 'loudest' man-made source in the ocean excluding large explosions). The sound travels through the seabed and reflects from the interfaces between the rock layers below. Very long (up to 10 km) arrays of hydrophones are towed near the surface to image the reflected sound, and the results are processed by computers to provide a three-dimensional map.

Though projectors are in principle just hydrophones in reverse, their physical design is often different. One of the most widely used types is the *Tonpilz* (German for 'singing mushroom') transducer, in which several piezoelectric PZT discs are sandwiched between electrodes to make a stack which is terminated by a conical or cylindrical *headmass*, the business end of the projector. Tonpilz transducers can generate frequencies in the 2 kHz to 50 kHz range.

For many underwater applications, including signalling and distance sensing, directional sounds are required. Just as in air, a sound source will be naturally directional if the sound waves it produces have smaller wavelengths than the transducer face. But because the velocity of sound is about five times greater in water than in air, the wavelength corresponding to a particular frequency is also about five times greater than its airborne equivalent, so directionality is harder to come by.

An elegant way to make a directional sound source is the parametric array. If two sound sources generate waves which differ just a little in frequency, then waves with that difference frequency will be produced, along with others whose frequency is the sum of those of the sources. The wavelength of the difference wave can be as long as required, but it maintains the directionality of its parent waves.

Parametric arrays exploit the fact that sound velocity depends on density. At high sound powers, the pressure in the compressions becomes very large, increasing density significantly and therefore briefly speeding up the sound wave; the reverse happens in the rarefactions. The effect of these velocity changes is to distort the wave from its usual sinusoidal form.

This is a common effect in high-power ultrasound, too. As Auguste Fourier showed, a non-sinusoidal wave is equivalent to a sum of component sinusoids. In the case of the parametric array, these components include the original waves, together with the sum and difference waves: the difference wave being the one of interest. Parametric arrays can also be used in air, to make audio-frequency sound more directional.

Although there is little that electromagnetic radiation does above water that sound cannot do below it, sound has one unavoidable disadvantage: its velocity in water is much lower than that of electromagnetic radiation in air, which means that scanning takes far longer. Also, when waves are used to send data, the rate of that data transmission is directly proportional to the wave frequency—and audio sound waves are around 1,000 times lower in frequency than radio waves. For this reason ultrasound is used instead, since its frequencies can match those of radio waves. Another advantage is that it is easier to produce directional beams at ultrasonic frequencies to send the signal in only the direction you want. However, a disadvantage is that absorption increases with sound frequency, so the range is limited.

Sounds worldwide

The distances over which sound can travel underwater are amazing. It is believed that before the proliferation of engine-powered vessels, Antarctic whales could be heard by their Arctic cousins. Such vast ranges are possible partly because sound waves are absorbed far less in water than in air. At 1 kHz, absorption is about

Box 11

Sound velocity in water: the most accurate (to 0.2 ms^{-1}) equation is the Leroy/Robinson formula: $c = 1402.5 + 5T - 0.0544T^2 + 0.00021T^3 + 1.33S - 0.0123ST + 0.000087ST^2 + 0.0156Z + 0.000000255Z^2 - 0.0000000000073Z^3 + 0.0000012(\Phi - 45) - 0.00000000000095TZ^3 + 0.0000003T^2Z + 0.0000143SZ$; T temperature (°C), S salinity (per cent), Z depth (metres), Φ latitude (degrees).

5 dB/km in air (at 30 per cent humidity) but only 0.06 dB/km in seawater. Also, underwater sound waves are much more confined; a noise made in mid-air spreads in all directions, but in the sea the bed and the surface limit vertical spreading.

The range of sound velocities underwater is far larger than in air, because of the enormous variations in density, which is affected by temperature, pressure, and salinity (Box 11). Levels at which sound velocity changes rapidly with temperature are called thermoclines, and they follow the same pattern in most seas: when the weather is calm, the uppermost layer of the sea is characterized by a rapid fall in temperature, and hence sound velocity, with depth. Because calm conditions are more common in summer, this is known as the seasonal thermocline. Below this is the main thermocline, where the temperature and sound velocity continue to fall with depth, independent of season. At the base of the main thermocline (the depth of which varies greatly with latitude), a steady temperature of around 4°C is reached and hardly changes at deeper levels. In this *deep isothermal layer*, pressure becomes the dominant factor in determining sound velocity—which consequently increases with depth, as shown in Figure 22.

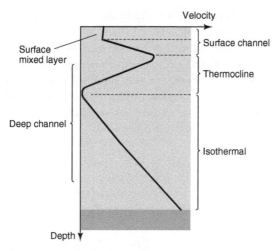

22. Underwater sound velocity profile.

So somewhere under all oceans there is a layer at which sound velocity is low, sandwiched between regions in which it is higher. By refraction, sound waves from both above and below are diverted towards the region of minimum sound velocity, and are trapped there. This is the deep sound channel, a thin spherical shell extending through the world's oceans.

Since sound waves in the deep sound channel can move only horizontally, their intensity falls in proportion only to the distance they travel, rather than to the square of the distance, as they would in air or in water at a single temperature (in other words, they spread out in circles, not spheres). Sound absorption in the deep sound channel is very low (it is strongly frequency dependent, but around 0.2 dB per km for 4 kHz waves), and sound waves in the deep channel can readily circumnavigate the Earth.

The deep sound channel was exploited to set up the SOFAR (sound fixing and ranging) system, which was initiated in 1960 by the Australia-Bermuda Sound Transmission Experiment, in which explosions were set off near Heard Island in the Indian Ocean off the coast of Australia. They were detected in Bermuda, at a distance of 20,000 km. A new, unexpected sound was discovered by the SOFAR researchers, and later identified as the calls of fin whales (*Balaenoptera physalus*), who long ago discovered the existence and properties of the deep sound channel and regularly visit it to signal their distant kin.

SOFAR opened the door to the ATOC (Acoustical Thermometry of Ocean Climate) system, which calculates global sea temperatures by measuring average sound velocity over large ranges, thus helping to quantify climate changes.

Changing weather causes changing sea conditions, which lead to a range of temporary acoustic states of the ocean, including *shadow zones* from which most sound is excluded, and temporary sound channels, which allow long-distance propagation. Sounds which travel down the latter, like those which travel around the deep sound layer, are highly distorted en route, due to the many changes of velocity on their journeys (due in turn to variations in temperature and salinity). In the 1990s, various strange sounds of unknown origin (the Bloop is perhaps the best known) led to a range of imaginative interpretations, including some literally monstrous ones. However, the most likely source is the greatly modified sounds of distant icebergs calving.

Earth sounds

Sound waves travel readily in solids (see Box 12), as do pressure waves of other kinds. But not all seismic waves are sound waves. P (for primary) waves are longitudinal: they are sequences of compressions and rarefactions whose velocity is determined by the density and elasticity of the ground, and therefore are sound waves. However, S (secondary) waves, being transverse, are not sound waves. Both P and S waves are *body waves*—they travel through the Earth, and their refraction by the layers underground provide us with information about our planet's structure. There are also a variety of surface seismic waves, but none are sound waves.

Many large animals make and hear low-frequency sounds. The sounds make by African elephants are low simply because their vocal cords are so large and hence relatively slow-moving—so much so that some of their calls are infrasonic. This is of considerable advantage to them since infrasound travels far with little attenuation (in air, very roughly, a 10 Hz signal travels one hundred times further than a 100 Hz one, and 10,000 times further than one at 1,000 Hz).

Underground, attenuation is highly variable, but usually much lower than in air: female elephants use infrasounds to attract males (over 3 km away by air or over 10 km underground) and to contact their young. Elephants also use infrasound to detect thunderstorms (a useful water source) over 500 km away. In 2004, Sri Lankan elephants fled the coast, probably because they detected the infrasounds of the oncoming tsunami. Elephants generate their signals either by making rumbling sounds or by stamping their feet, which also detect ground-borne sound using vibration receptors called Pacinian corpuscles.

As far as humans are concerned, the use of underground sound goes back at least to 132 CE, when the first seismic detector was made in China. It was a brass vessel with a ring of metal frogs attached, each holding a metal sphere in its mouth. The presence

Box 12

Attenuation of sound in gas or liquid media is described by Stokes' Law: $\alpha = 2\eta f^2 / 3\rho v^3$; amplitude α, dynamic viscosity η (nu), frequency f, density ρ (rho), velocity of sound v.

and bearing of an earthquake was indicated by the fall of the sphere(s) in its direction.

Underground sound has been a source of fear and wonder since ancient times, reactions that evolution may have hardwired into our brains, from the experiences of our ancestors that such sounds accompany avalanches, volcanic eruptions, earthquakes, and other overwhelming natural disasters. The fact that they can be felt as well as heard adds to their impact: to feel the solid earth shift and tremble beneath one's feet is an unnerving experience. No wonder that the dark world of the underground was regarded in many cultures as the abode of the dead.

Despite its unwholesome reputation however, underground sound has long been used in war: during classical times, tunnelling enemies were sometimes detected by the sounds they made through the earth, and there is even a record of shields being clashed on to the ground and the presence of tunnels below being judged by the sounds heard. In World War I, detections from different points along trench walls allowed the locations of enemy troops in their own trenches to be estimated, by triangulation. In World War II, the Polish resistance spent significant periods underground, listening for and tracking German soldiers overhead—who were meanwhile tracking them by the same means.

Underground sound detectors are called geophones. Until the late 20th century, geophones worked on the same principle as dynamic microphones (Chapter 5): a magnet was set in motion by ground waves and induced electrical signals in a wire coiled around it. Today, MEMS are used instead: a microscopic piece of silicon is mounted in a delicate holder and vibrations due to underground sound cause it to start to move. A feedback system halts this motion and the force it applies to do so provides precise information about the sound. However, MEMS devices are relatively insensitive and are used mainly for monitoring actively seismic regions. All practical geophones are highly directional and are usually deployed to respond to sounds coming directly from below.

Prospecting

One well-established use of underground sound is prospecting. Usually an explosive charge is set off underground (there is no underground equivalent to the loudspeaker), and the waves it makes reflect off the interfaces between layers of different materials and are detected by an array of geophones.

The measurement and localization of strain-produced ultrasounds in solids, known as acoustic emission, is used to detect the onset of fractures in all sorts of structures, from planes in flight to the London Eye. This is known as structural health monitoring. Alerts can be triggered at the instant that even an invisibly small crack begins to form. Acoustic emission is also used to study the formation of cracks during welding processes, to detect the onset of lesions in pipes that carry high-pressure fluids and to determine the amount of corrosion inside reinforced concrete.

Underground nuclear tests, difficult to identify by other means, generate distinct sounds which are mostly infrasonic. The CTBTO (Comprehensive Nuclear-Test-Ban Treaty Organization) constantly monitors the Earth using geophones (together with hydrophones in the deep sound channel to detect underwater sounds originating from the sub-ocean rocks). This network of underground and underwater detectors allows the CTBTO to police international test ban treaties, and to determine the locations of any tests that do take place.

This chapter has focussed on those areas of sound that we can't hear, even though many scientists wish we could—in Chapter 8 we turn to sounds that we cannot help hearing, even though we'd rather not.

Chapter 8
Sound out of place

The nature of noise

What is noise? This is one of those questions like 'What is time?', to which one might answer: 'I know exactly what it is until someone asks me'. But it's worse than that, for 'noise' means two quite different things. For a scientist, noise is extraneous acoustic or electromagnetic energy. When used for communication, noise is whatever is *not* the signal, in other words does not carry information (whether the information is a voice down a wire, or a structural element in an ultrasonically scanned foetus). The larger the amount of noise, the more difficult it becomes to unweave the signal from it—hence the concept of signal to noise ratio.

Once such noise has been identified, it can be fairly easy to remove. Noise in a circuit that occurs at particular frequencies (a 50 Hz hum from a mains electricity supply, for example) can often be dealt with by filters, so that only frequencies of interest remain. Unfortunately, in electronics, a very common form of noise is white noise, a signal that fills all frequencies of interest and sounds like a hiss when heard through a loudspeaker. This cannot be filtered out so easily, but one may instead filter all but the frequency bands that carry the signal; there will still be white noise in those bands, but the overall signal to noise ratio will improve.

For most of us most of the time though, 'noise' is *any* sound which is unwanted by the person who is exposed to it. A trumpeter may be very pleased with his output and certainly not think of it as noise, but a neighbour might well disagree, however technically excellent it may be. And both are right—one person's music is another's noise. This disagreement goes to the very heart of why noise is so difficult to quash. But fortunately, while almost every sound has at least a few people who love it, and there are those to whom almost anything but silence is anathema, it is also true that sounds which many of us think of as noise do tend to share three characteristics: suddenness, loudness, and tunelessness. To find out why this is so, we need to consider why we can hear anything at all.

Ancestral legacies

We hear because our ancestors did, and they heard because their deaf relatives died. Among the many sounds with which they would have been surrounded, a few were matters of life and death. A sudden roar, the twang of a bow string, the snap of twigs underfoot, the crack of thunder overhead were all danger signs, and what they have in common is suddenness—hence today our instant reactions to sudden noise. Sudden silence, like the hush of bird song when a predator is sighted, meant as much as sudden sound and hence causes the familiar deafening silence when a clock stops ticking, an air conditioner switches off, or persistent rain suddenly ceases. In fact, *any* unexpected change to the sounds around us can be an annoying noise: every driver knows the irritation and concern caused by a new and unidentifiable engine noise, which an unsympathetic passenger may not be able to pick out at all, however well the driver imitates it.

But why are we annoyed by those sounds which seem unrelated to any ancestral sound that might signal danger? People talking in the street outside at midnight? MP3 players you can just hear across a train carriage?

The answer lies in a key social function of sound: to demonstrate power, in particular by claiming ownership of space. We are used to the idea that each of us has a personal bubble of space that we carry round, to which the uninvited are unwelcome. Someone who invades that space with their sounds can be as annoying as someone intruding into it physically, and the main point in both examples is intent. How much more annoying is the person sitting next to you if they phone someone to talk nonsense than if they receive a call of real concern?

Whether another person is regarded as noisy or not depends too on the relationship between them and the listener. Historians Shane and Graham White, for instance, in their studies of slavery in the US, found that the sermons of African American evangelist preachers, while appreciated by congregations, were referred to simply as noisy by white American Christians who were concerned because the sounds, and perhaps their messages too, spread far beyond the confines of their venues. Cultural norms are important too: acceptable sound levels in libraries, during mealtimes, and at funerals vary greatly in different countries.

The sea of other people's noise in which one is often submerged on public transport, in public spaces, or open-plan offices can be a disturbing environment. The individual sounds can be neither identified nor located, but the hearing system constantly tries to do both, making them as hard to ignore as to escape. Even when one does manage to ignore such sounds on a conscious level, the mental processing continues, as does the stress that sometimes results—and the high blood pressure that this stress can cause.

Often, the only way to reclaim one's private sound bubble is to fill it with a sound field of one's own making, by use of an MP3 player or smartphone. According to Michael Bull, Professor of Media and Film at the University of Sussex (UK): 'iPod use can usefully be interpreted as a form of pleasurable toxicity within which the "total mediated" world of users lies a dream of unmediated experience—of direct access to the world and one's emotions'.

There are some sounds which cannot be ignored, even though they do not really exist. These so-called *earworms* are usually brief snatches of music, often banal ones, which insist on repeating themselves in the mind. Advertising jingles are especially good at worming their way in. Though so widespread and seemingly simple, earworms are actually very hard to explain. Neurologist Oliver Sacks suggests that the unusual way music is remembered may be significant. Recalling a scene or event involves reconstructing it, which in turn means that the result may be very different in different 'replays'. In recalling a piece of music, however, something much closer to a direct copy of the original is preserved, an 'almost defenceless engraving of music on the brain', as Sacks puts it, over which we have relatively little control. Perhaps the fact that tunes, and hence this 'engraving' process, arrived so late in the evolution of the mind is a reason why such memories are hard to control.

Music can have even more profound effects on the mind—in some people it can cause epileptic seizures, while in others it can have a significantly calming effect, and can even alleviate pain and high blood pressure. It can be especially effective in the treatment of some mental problems, and has been used in this way since World War II. Music has also been shown to be of great relief to some who suffer from Parkinson's disease.

Deafening

The problem—indeed the tragedy—of noise-induced hearing loss is that it may not happen until years after the damage has been done. Also, while we react immediately to dangerously high temperatures or intense light, our defensive reactions to noise are far poorer.

Our eyes are equipped with a range of protective adaptations, most obviously eye lids and the contracting pupil (or expanding iris). Why does the ear have neither? The reason for the lack of ear lids is that while deafness may shorten an animal's life, it won't shorten it anything like as much as being successfully crept up on by a hungry tiger. And we do have an equivalent to the pupil, though it's not particularly effective: the acoustic reflex is provided by two muscles in each ear called the stapedius and the tensor tympani. When a loud sound arrives, the stapedius pulls the ossicle called the stapes ('stirrup', named from its shape) away from the oval window, while the tensor tympani pulls on the malleus (hammer) ossicle, which is attached to the eardrum and hence stiffens the latter. As a result, sounds are muffled. The stapedius usually tenses when we speak, to stop us annoying ourselves with our own voices; the tensor tympanum does the same job when we eat, to suppress our own chewing noises.

The acoustic reflex is one cause of temporary threshold shift (TTS), in which sounds which are usually quiet become inaudible. Unfortunately, the time the reflex takes to work (called its latency) is usually around 45 milliseconds, which is far longer than it takes an *impulse sound*, like a gunshot or explosion, to do considerable damage. The reflex is only one contributor to TTS; the other mechanisms are still unknown.

Where the overburdened ear differs from other abused measuring instruments (biological and technological) is that it is not only the SPL of noise that matters: energy counts too. A noise at a level which would cause no more than irritation if listened to for a second can lead to significant hearing loss if it continues for an hour. The amount of TTS is proportional to the logarithm of the time for which the noise has been present—that is, doubling the exposure time more than doubles the amount.

However, very loud impulse sounds, like gunshots, will also cause instant damage, even if the total energy is low. (Reputedly, a trained audiologist can identify the permanent damage caused by a single shot fired by a patient with no ear protection.) Recovery times increase with the amount of shift; a TTS of 40 dB can take weeks to recover from. The amount of TTS reduces considerably if there is a pause in the noise, so if exposure to noise for long periods is unavoidable (at a football match, say), there is very significant benefit in removing oneself from the noisy area, if only for fifteen minutes.

It is on the evidence of many careful experiments on the effects of noise on hearing that noise regulations, now statutory in workplaces and elsewhere in many countries, have been defined. They place limits both on peak noise levels and durations of exposure to specified lower levels.

One problem with identifying how many people have noise-induced hearing loss is that everyone's ability to hear high-frequency sounds declines with age: newborn, we can hear up to 20 kHz, by the age of about forty this has fallen to around 16 kHz, and to 10 kHz by age sixty. Aged eighty, most of us are deaf to sounds above 8 kHz. The effect is called presbyacusis (literally 'old man hearing'). Since noise-induced hearing loss usually also impacts higher frequencies and, being cumulative, becomes more common with age, we don't really know how much loss is avoidable.

If there has been a downside to the often laudable progress made in tackling dangerously high levels of noise, it has been the relative neglect of noise which is 'only' loud enough to annoy people, despite their vast numbers: the 2008 Ipsos MORI (UK) *National Noise Survey* found that 26 per cent of respondents were annoyed by neighbour noise, for example. Those affected lose sleep, they lose concentration, they lose patience—they lose the joy from their lives (39 per cent said their quality of life was adversely affected). Such noise also exacerbates mental health problems, community tensions, and social isolation.

There are few who would deny that the best approach would be to remove noise from the environment in the first place. How this can be done, and how effectively, depends crucially on the source. To most inhabitants of the developed world and many elsewhere, the following sources are likely to be the main ones:

- Air traffic
- Industrial
- Neighbour
- Neighbourhood
- Rail
- Road traffic

In addition, shipping noise and wind-farm noise affect some areas significantly, while most areas are blighted by construction noise from time to time.

Of these groups, neighbour and neighbourhood noise are particularly difficult to deal with, partly because the maker does not class them as noise but the listener does. This difficulty has led sound historian Karin Bijsterveld to conclude that noise abatement is in the grip of a 'paradox of control', in which the promises of experts and politicians to control noise actually amount to passing the buck to neighbours to do so, or developing such complex formulae for limiting other forms of noise, such as that from aircraft, that few can understand, engage with, or critique them.

Fighting back

With the exception of some neighbour noise and some warning signals, no one would be likely to complain if noise were to be eliminated. Can this be done?

In those (unfortunately rare) cases in which noise is highly regular, consists of one or few frequencies, and is bothersome only in a small and well-defined location, active noise cancellation (ANC) can be highly effective. ANC relies on the fact that sound waves consist of compressions followed by rarefactions. On a pressure plot such as Figure 1, this is clear from the fact that the line goes first above and then below the midpoint. If a second sound wave were superimposed on the first such that at every point at which there is a compression in the original wave there was an equally intense rarefaction in the new one, perfect silence would result, the acoustic energy being converted entirely to heat. However, in practice this can only be achieved over very limited areas, such as plane cockpits, the driver's head area of some cars and—perhaps most usefully—the space inside ear defenders (Figure 23).

Many highly effective technological solutions to noise have been developed. The problem is that all of them have been applied already. Today's vehicles are extremely quiet compared to their

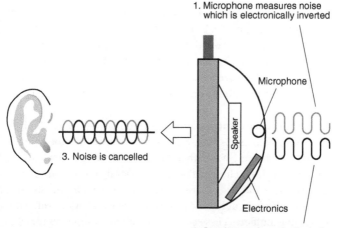

1. Microphone measures noise which is electronically inverted

Microphone

Speaker

3. Noise is cancelled

Electronics

2. Speaker produces 'antinoise', identical in level and waveform but 180° out of phase

23. Active noise cancellation.

mechanical powers, far quieter than their much feebler ancestors ever were, but this fact is overwhelmed by the vastly greater number of them in use today. But it is still worth looking at how today's machines are kept so quiet, since the same principles will almost certainly be applicable in future, especially as new materials become available.

The first principle of noise control is to identify the source and remove it. For perhaps the single most important noise source today, the internal combustion engine (ICE), this is impossible: the noise is made by the explosions within, which are intrinsically noisy. However, as explained in Chapter 3, our hearing systems do not measure the total energy of a sound—they are tuned to respond most strongly to frequencies around 4 kHz, and it is usually possible to shift the output frequency of ICEs to below this. The main means are the adjustment of the timing and the rate of fuel injection, to modify the combustion of the fuel so that such high-frequency components are reduced.

Having dealt as far as possible with the noise source, the next step is to contain it. Again, ICEs are challenging since they have to be connected to the air around them, to draw in oxygen and expel exhaust gases, and air paths are noise routes. To reduce noise emissions, mufflers are the answer. (A silencer is really the same thing, but is more often so-called when used to quieten firearms.)

There are two basic muffler principles, and most ICEs use both. Absorptive silencing is achieved simply by lining the pipe with absorber so that any waves that do not travel straight down are converted to heat. Those which do travel straight down are also reduced, since their edges are retarded by the duct lining. But this effect is only slight at low frequencies, which are better dealt with by reactive mufflers. These contain Helmholtz resonators tuned to the most troublesome frequencies. Sounds at these frequencies are thus enhanced inside the muffler, allowing their energies to be absorbed there. A few mufflers also use ANC.

Over the next few decades, it seems likely that ICEs will largely give way to electric versions. There is no technical reason why these should not be almost silent, but a vehicle which approaches silently could be deadly, so all will be provided with specially generated running and warning sounds (which might also be of value for bicycles—especially those with no bells). There is no reason for these to be unpleasant, however—even an alarm need not be alarming. To attract attention, 'shushing' sounds can do the same job with less annoyance.

Jet engines, on the other hand, are probably here to stay for the foreseeable future. In their case, the source of the noise is not the burning gases themselves, but what happens when they emerge from the engine at high speed and mix with the relatively slow-moving air outside the plane. The turbulence this causes is the source of almost all the noise. Enormous efforts have been made to smooth the mixing, mainly by increasing the circumference of the jet by corrugating it and by cocooning the exhaust jet in air of intermediate speed. No more can be done in these directions without an excessive loss of thrust, but other approaches have been considered. One idea that has met with modest success is to mount engines above the wings. Another is a strategic change, such as the use of airships for the transport of goods where transit times are not an issue.

All but the slowest or most antiquated moving vehicles are now streamlined, which avoids the turbulences that would produce noise as well as increasing efficiency. Streamlining is achieved partly by designing the vehicle with a suitable shape so that the airstream slips past it without the sudden change of direction or speed that would result in turbulence, and partly by smooth surfacing. Nonetheless, as far as car drivers are concerned, the aerodynamic noise which streamlining is intended to deal with is still the most significant at cruising and higher speeds. The principal sources are usually the wing mirrors and the A pillars (the ones which hold the sides of the windscreen in place).

For those on the pavement, the loudest noise is frequently made by the tyres, especially on concrete roads. Low-noise surfaces can do a great deal to reduce this. The best available are layers of asphalt with many air-filled pores (about 25 per cent of volume), but such surfaces are costly and not very robust. A less effective, but more durable and cheaper alternative, is the use of thin (~2 cm) asphalt layers.

Time was when the design of machines did not consider quietness at all, and noise management was applied only as a sticking plaster afterwards or as unreliable rules of thumb (like stretching a few strings under concert hall floors or furnishing theatre stages with vases). Now, machines are usually well designed and carefully engineered to be at least fairly quiet—though regular maintenance of industrial machinery and careful installation of domestic appliances are essential.

When noise can be neither avoided not contained, the next step is to keep its sources well separated from potential sufferers. One approach, used for thousands of years, is zoning: legislating for the restriction of noisy activities to particular areas, such as industrial zones, which are distant from residential districts. The first recorded example dates from around 700 BCE, when tinsmiths, blacksmiths, carpenters, potters, and even roosters were banned from the city centre of Sybaris, a Greek colony on the Aegean coast.

Where zone separation by distance is impracticable due to historical or geographical considerations, sound barriers are the main solution: a barrier that just cuts off the sight of a noise source will reduce the noise level by about 5 dB, and each additional metre will provide about an extra 1.5 dB reduction. Also, the barrier must be massive enough to prevent much sound passing straight through it; about 10 kg under each square metre is usually sufficient. Since barriers largely reflect rather than absorb, reflected sounds need consideration, but otherwise design and construction are simple, results are predictable, and costs are relatively low.

Temporary flexible versions of noise barriers, known as acoustic fencing, can be placed around construction sites and, if carefully positioned and arranged to avoid leaving gaps, can be very effective, reducing noise levels by up to 30 dB in some cases. Unlike permanent noise barriers, acoustic fencing achieves noise reduction primarily by absorption.

The big problem with noise barriers is their unattractiveness to the eye. This can be offset by introducing a screen of vegetation between barrier and observer. If space permits at least 10 metres of vegetation, the plants will help reduce the noise somewhat too. At low frequencies (around 25 Hz), this is mostly due to the extra ground absorption caused by fallen leaves, and at high frequencies (above ~1 kHz) it is caused by foliage. It is best if the leaves are half as long as the sound waves, so, for both these reasons, some deciduous content is essential, (a mix of trees and bushes) reaching as high as possible (up to about 10 metres, after which reflection from large branches reduces the effect). On windy days, the sound of such vegetation is another bonus, as a distraction from the noise.

The addition of natural sound sources (soundscaping) is especially effective in urban parks. In such spaces the commonest way to improve a soundscape is by adding fountains, but birdsong (real or recorded) can work well too.

Quietening homes

Of all spaces, the most precious is home, because we relax and sleep there, and because it is an extension of our personal space. Keeping noise out is, therefore, a high priority: the best way to do this is through good acoustic design, including selection of the right materials, and, crucially, their proper installation and maintenance. This is much cheaper, more aesthetically pleasing, and more successful that retrofitting noise solutions into a pre-existing dwelling.

Whether retrofitted or not, the basic approaches to home sound reduction are simple: stop noise entering, destroy what does get in, and don't add more to it yourself. There are three ways for sound to enter: via openings; by structure-borne vibration; and through walls, windows, doors, ceilings, and floors acting as diaphragms. In all three cases, the main point to bear in mind is that an acoustic shell is only as good as its weakest part: just as even a small hole in an otherwise watertight ship's hull renders the rest useless, so does a single open window in a double-glazed house. In fact, the situation with noise is much worse than with water due to the logarithmic response of our ears: if we seal one of two identical holes in a boat we will halve the inflow. If we close one of two identical windows into a house, the inflow will again half—but that 50 per cent reduction in acoustic intensity is only about a 2 per cent reduction in loudness.

The second way to keep noise out is double glazing, since single-glazed windows make excellent diaphragms. Structure-borne sound is a much greater challenge, though if the source is within the dwelling, the room which it (or (s)he) occupies, can be treated. But this will not be cheap and will usually require treatment of the ceiling as well as all walls, the floor, and the door.

One inexpensive, adaptable, and effective solution—though maybe not the most attractive—is the hanging of heavy velour drapes, with as many folds as possible. If something more drastic is required, it is vital to involve an expert: while an obvious solution is to thicken walls, it's important to bear in mind that doubling thickness reduces transmission loss by only 6 dB (a sound power reduction of about three-quarters, but a loudness reduction of only about 40 per cent). This means that solid walls need to be very thick to work well.

A far better approach is the use of porous absorbers and of multi-layer constructions. In a porous absorber like glass fibre, higher-frequency sound waves are lost through multiple reflections from the many internal surfaces. Using stud-mounted panels with air gaps between them and the wall will deal with lower frequencies, through reflection of lower frequency waves at the air/solid interface (yet another example of an impedance mismatch). A well-fitted acoustically insulated door is also vital.

The floor should not be neglected: even if there are no rooms beneath, hard floors are excellent both at generating noise when walked on and in transmitting that noise throughout the building. Carpet and underlay are highly effective at high frequencies but are almost useless at lower ones, so if you have floor-mounted loudspeakers or an extrovert baritone in the home, something special will be required, and again there is no real alternative to bringing in an expert.

Though rarely a problem in the home, the most significant issue in the design of public spaces is reverberation. Surprisingly, though it has been a bugbear for architects, performers, and audiences alike for millennia, no way to quantify it was developed until 1898, when Wallace Clement Sabine defined the reverberation time of a room as the period necessary for the intensity of the sound to decline to one millionth (–60 dB) of its initial level (Box 13). Even more usefully, Sabine derived an empirically based equation that allowed this time to be calculated from the size and shape of a room—so architects could predict it at last.

Box 13

Sabine's equation: $T = 0.161 (V/A)$, reverberation time in seconds T, room volume V, total absorption area A. This equation is still used today except for highly absorbent rooms.

However much care, money, and expertise is spent on an individual noise problem, the results will be ineffective unless both the root cause and the context of that problem are investigated first: it may be pointless to replace standard doors with acoustic ones if the actual problem is that they are routinely left open, and there is little point in reducing the speed limit on a trunk road if the effect is to send traffic down residential streets. Such holistic approaches to noise reduction extend beyond acoustic treatments: the benefits of fitting sound absorbing panels must be balanced against their visual impact, the introduction of artificial sounds to distract park users from traffic noise may annoy more people than it pleases, and moving people from open-plan offices to quieter cellular ones may reduce team working. Similarly, while well-built houses can keep noise out, this necessarily involves keeping the inhabitants in, cutting off the sounds of nature too, and leading to alienation. As David Hendy puts it: 'whenever we withdraw into separate soundscapes...we make strangers of each other'.

The wider view

Despite the great importance to us of our hearing systems, their performance is not usually a matter of life and death. This is not the case for some underwater species. The prime importance of sound to many marine animals, combined with the efficiency with which sounds travel underwater, means that some marine creatures are devastated by the effects of noise, in particular whales and dolphins. The scale of the effects is unclear but some are subtle: since whales communicate with their calves by sound signals, the mother/child link can easily be broken by extraneous sounds and the calf separated. In other cases, the startle effect of unfamiliar loud sounds on whales and dolphins can be so extreme that rapid surfacing occurs, leading to death from the formation of nitrogen gas bubbles in the liver and other organs (the condition divers describe as 'the bends').

There are many contributors to underwater noise pollution. As well as the sounds of shipping, sonar systems, blasting, and signalling all add their load. Fortunately, there is now widespread recognition of the effects on animals; more in fact than on divers, who routinely risk severe hearing damage (although, as the effects are often the same as those of pressure changes in the ear, the scale of the problem is hard to establish).

The interdisciplinary and public nature of attempts to battle noise, both on land and underwater, make collaboration at all levels vital, and cannot succeed without the active support of international organizations such as the World Health Organization, national governments, local authorities, specialist organizations like the UK Noise Abatement Society, and the public. One activity which should bring all these players together is noise action and awareness weeks.

As technologies have evolved, cities have grown, and vehicular travel has increased, noise sources have proliferated and noise is now a fact of life for many, whether made deliberately and selfishly via loudspeakers or unavoidably and carelessly through engines and other hardware. Although there is much that science can do to help, it must be accompanied by both education and legislation which ensures that noise is universally regarded as the pollutant that it is.

Ultimately, noise-making must become unacceptable, rather than just annoying. While this may seem impractical, exactly such a change has taken place in the UK regarding smoking—and this took only a few years. The change came about through a combination of new laws implemented through both national and local government, and a loud and clear message that smoking causes serious illness. Given the same approach to noise, just as radical a change might be accomplished.

* * *

Of all disciplines, sound has by far the widest variety of interested parties. Understanding it enhances the work of actors, advertisers, aerospace engineers, automotive engineers, anthropologists, architects, artists, broadcasters, builders, communication engineers, composers, designers, ecologists, educators, electronic engineers, environmental health officers, film-makers, historians, marine biologists, musicians, physicians, physicists, politicians, prospectors, psychologists, seismologists, sociologists, town planners, zoologists, and many others.

Hitherto, many of these domains have been separate, but over the past few decades approaches like those of soundscaping, sound studies, and the design of ever more intimate and personalized media systems have started to bring together expertise from across these fields. Such collaborations are often hampered by the narrow training offered to most practitioners, but meanwhile they are making people more aware of the vast range and power of the subject—of the multitude of practical and emotional benefits a knowledge of sound can lead to, and of the many new applications and insights the bringing together of its many branches may yield. Such developments are to the benefit of everyone. For all of us, the future sounds good.

References

Chapter 1: Past sounds

Details of the sound of the early universe may be found here: <http://www.astro.virginia.edu/~dmw8f/BBA_web/index_frames.html>; the history of acoustics is detailed in Dayton Clarence Miller, *Anecdotal History of the Science of Sound to the Beginning of the 20th Century* (Macmillan, 1935) and Robert T. Beyer, *Sounds of Our Times: Two Hundred Years of Acoustics* (Springer, 1999).

Further historical material is in: David Hendy, *Noise: A Human History of Sound and Listening* (Profile Books, 2014). The acoustics of Epidaurus are explained at: <http://www.nature.com/news/2007/070319/full/news070319-16.html>. The most recent edition of R. Murray Schafer's classic book on soundscapes is *Soundscape: Our Sonic Environment and the Tuning of the World* (Destiny Books, 1994). The Attali material is from Mark M. Smith (ed.) *Hearing History: A Reader* (University of Georgia Press, 2004). Alain Corbin's classic work (translated by Martin Thom), is *Village Bells: The Culture of the Senses in the Nineteenth-Century French Countryside* (Columbia University Press, 1998).

'simultaneously a physical environment and a way of perceiving that environment', Emily Thompson, *The Soundscape of Modernity: Architectural Acoustics and the Culture of Listening in America, 1900–1933* (MIT Press, 2002), p. 1.

Chapter 2: The nature of sound

Most of the material here may be found in Thomas D. Rossing, Richard F. Moore, and Paul A. Wheeler, *The Science of Sound* (Addison Wesley, 2001) and Daniel R. Raichel, *The Science and Applications of Acoustics* (Springer, 2006). Most cultural references are from Jonathan Sterne, *The Sound Studies Reader* (Routledge, 2012).

Chapter 3: Sounds in harmony

Material from this chapter is from Neville H. Fletcher and Thomas D. Rossing, *The Physics of Musical Instruments* (Springer, 2008), Daniel J. Levitin, *This is Your Brain on Music: Understanding a Human Obsession* (Atlantic Books, 2008), John Powell, *How Music Works: A Listener's Guide to Harmony, Keys, Broken Chords, Perfect Pitch and the Secrets of a Good Tune* (Particular Books, 2010), and Eric Taylor, *The AB Guide to Music Theory* (Oxford University Press, 2013).

Chapter 4: Hearing sound

Information in this chapter is largely from David Howard and Jamie Angus, *Acoustics and Psychoacoustics* (Focal Press, 2000), Lawrence J. Raphael, Gloria J. Borden, and Katherine S. Harris, *Speech Science Primer* (Lippincot Williams & Wilkins, 2003), William Yost, *Fundamentals of Hearing: An Introduction* (BRILL, 2013), and H. Zwicker and H. Fastl, *Psychoacoustics: Facts and Models* (Springer, 2006). The Barthes material is from Bruce R. Smith's *The Acoustic World of Early Modern England: Attending to the O-Factor* (University of Chicago Press, 2nd edition, 1999).

'My feeling is that an entire culture...', Brandon LaBelle, *Acoustic Territories: Sound Culture and Everyday Life* (Continuum, 2010), p. xvi.

'the rich undulations of auditory material...', LaBelle, *Acoustic Territories*, p. xxi.

'Sound creates a relational geography...', LaBelle, *Acoustic Territories*, p. xxv.

The statistics on worldwide hearing loss are from the World Health Organization's 'Deafness and Hearing Loss' fact sheet (number 300), Updated March 2015. It can be found at <http://www.who.int/mediacentre/factsheets/fs300/en/>.

Chapter 5: Electronic sound

Technological details are to be found in Glen Ballou, *Electroacoustic Devices: Microphones and Loudspeakers* (Focal Press, 2009), F. Alton Everest and Ken C. Pohlmann, *Master Handbook of Acoustics* (Tab Electronics, 2009). This book, and Bruce and Marty Fries, *Digital Audio Essentials* (O'Reilly Media, 2005), also give practical guidance, while historical material is from Greg Milner, *Perfecting Sound Forever: The Story of Recorded Music* (Granta, 2010) and Roland Gelatt, *The Fabulous Phonograph, 1877–1977* (Cassell, 2nd edition, 1977).

Material on the effects of recordings on performance are from Robert Philip, *Performing Music in the Age of Recording* (Yale University Press, 2004) and Mark Katz, *Capturing Sound: How Technology Has Changed Music* (California University Press, 2010).

'the MP3 carries within it practical and philosophical understandings…', Jonathan Sterne, *MP3: The Meaning of a Format (Sign, Storage, Transmission)* (Duke University Press, 2012), p. 2.

Chapter 6: Ultrasound and infrasound

Material from this chapter is from Vivien Gibbs, *Ultrasound Physics and Technology: How, Why and When* (Churchill Livingstone, 2009), Dale Ensminger and Leonard J. Bond, *Ultrasonics: Fundamentals, Technologies, and Applications* (CRC Press, 2011), Tim Leighton, *The Acoustic Bubble* (Academic Press, 1994), and Gillian Sales and David Pye, *Ultrasonic Communication by Animals* (Chapman & Hall, 1974). The 2015 bat study referred to in this chapter is Pavel Kounitsky et al., 'Bats Adjust their Mouth Gape to Zoom their Biosonar Field of View', *Proceedings of the National Academy of Sciences* 2015, Vol. 112, No. 21, pp. 6724–9.

Chapter 7: Sound underwater and underground

Most technological and physics material is from Xavier Lurton, *An Introduction to Underwater Acoustics: Principles and Applications* (Springer Praxis Books, 2010), M. A. Ainslie, *Principles of Sonar Performance Modelling* (Springer Praxis Publishing, 2010), and Robert J. Urick, *Principles of Underwater Sound* (Peninsula Publishing, 2013). Biological content is from W. J. Richardson,

C. R. Greene Jr, C. I. Malme, and D. H. Thomson, *Marine Mammals and Noise* (Academic Press, 1995). Historical material is from Robert T. Beyer, *Sounds of Our Times: Two Hundred Years of Acoustics* (Springer, 1999).

Chapter 8: Sound out of place

Most material is from Mike Goldsmith, *Discord: The Story of Noise* (Oxford University Press, 2012). Cultural references are from Smith and Karin Bijsterveld, *Mechanical Sound: Technology, Culture, and Public Problems of Noise in the Twentieth Century* (MIT Press, 2008).

'iPod use can usefully be interpreted...', Michael Bull, 'iPod: A Personalized Sound World for its Consumers', *Revista Comunicar*, 2010, Vol. XVII, No. 34, pp. 55–63.

Oliver Sacks' account of earworms is from his book *Musicophilia* (Picador, 2011), p. 47.

'whenever we withdraw into separate soundscapes...we make strangers of each other', James Hendy, *Noise: A Human History of Sound and Listening* (Profile, 2013), p. 331.

Further reading

Chapter 1: Past sounds

Robert T. Beyer, *Sounds of Our Times: Two Hundred Years of Acoustics* (Springer, 1999).

David Hendy, *Noise: A Human History of Sound and Listening* (Profile Books, 2014).

Mark M. Smith (ed.), *Hearing History: A Reader* (University of Georgia Press, 2004).

Chapter 2: The nature of sound

Trevor Cox, *Sonic Wonderland: A Scientific Odyssey of Sound* (The Bodley Head, 2014).

Trevor Pinch and Karin Bijsterveld, *The Oxford Handbook of Sound Studies* (Oxford University Press, 2013).

Daniel R. Raichel, *The Science and Applications of Acoustics* (Springer, 2006).

Thomas D. Rossing, Richard F. Moore, and Paul A. Wheeler, *The Science of Sound* (Addison Wesley, 2001).

Jonathan Sterne, *The Sound Studies Reader* (Routledge, 2012).

Chapter 3: Sounds in harmony

Daniel J. Levitin, *This is Your Brain on Music: Understanding a Human Obsession* (Atlantic Books, 2008).

John Powell, *How Music Works: A Listener's Guide to Harmony, Keys, Broken Chords, Perfect Pitch and the Secrets of a Good Tune* (Particular Books, 2010).

Chapter 4: Hearing sound

David Howard and Jamie Angus, *Acoustics and Psychoacoustics* (Focal Press, 2000).

William Yost, *Fundamentals of Hearing: An Introduction* (BRILL, 2013).

Chapter 5: Electronic sound

F. Alton Everest and Ken C. Pohlmann, *Master Handbook of Acoustics* (Tab Electronics, 2009).

Roland Gelatt, *The Fabulous Phonograph, 1877–1977* (Cassell, 2nd edition, 1977).

Greg Milner, *Perfecting Sound Forever: The Story of Recorded Music* (Granta, 2010).

Chapter 6: Ultrasound and infrasound

Dale Ensminger and Leonard J. Bond, *Ultrasonics: Fundamentals, Technologies, and Applications* (CRC Press, 2011).

Vivien Gibbs, *Ultrasound Physics and Technology: How, Why and When* (Churchill Livingstone, 2009).

Tim Leighton, *The Acoustic Bubble* (Academic Press, 1994).

Chapter 7: Sound underwater and underground

Xavier Lurton, *An Introduction to Underwater Acoustics: Principles and Applications* (Springer Praxis Books, 2010).

Robert J. Urick, *Principles of Underwater Sound* (Peninsula Publishing, 2013).

Chapter 8: Sound out of place

Mike Goldsmith, *Discord: The Story of Noise* (Oxford University Press, 2012).

Garrett Keizer, *The Unwanted Sound of Everything We Want* (Public Affairs, 2012).

Oliver Sacks, *Musicophilia* (Picador, 2011).

John Stewart, Francis McManus, Nigel Rodgers, Val Weedon, and Arline Bronzaft, *Why Noise Matters: A Worldwide Perspective on the Problems, Policies and Solutions* (Routledge, 2011).

Index

Sound

SOCIAL MEDIA
Very Short Introduction

Join our community
www.oup.com/vsi

- Join us online at the official Very Short Introductions **Facebook** page.
- Access the thoughts and musings of our authors with our online **blog**.
- Sign up for our monthly **e-newsletter** to receive information on all new titles publishing that month.
- Browse the full range of Very Short Introductions online.
- Read **extracts** from the Introductions for free.
- Visit our library of **Reading Guides**. These guides, written by our expert authors will help you to question again, why you think what you think.
- If you are a teacher or lecturer you can order inspection copies quickly and simply via our website.

ONLINE CATALOGUE
A Very Short Introduction

Our online catalogue is designed to make it easy to find your ideal Very Short Introduction. View the entire collection by subject area, watch author videos, read sample chapters, and download reading guides.

http://global.oup.com/uk/academic/general/vsi_list/